Ancestral Nutrition

A Modern Revival with Peptides

Peptides: Fueling the Very Roots of Every Cell, Empowering
Life at its Most Fundamental Level

Ancestral Nutrition
A Modern Revival with Peptides

Copyright © *Levitas One*, 2024
All Rights Reserved

What are the NoMAD Plans?

Developed by Dr Ash Kapoor, the NoMAD Plans represent a transformative approach to health and wellness that combines the wisdom of ancestral practices with contemporary medical insights. The name "NoMAD" not only suggests a journey through the intricate realm of health but also stands for its foundational principles: Nutritional Optimisation, Mindful Adaptation, and Detoxification.

At the heart of NoMAD is the 6 R Framework—Restore, Release, Repair, Renew, Reframe, and Represent. This methodology addresses the root causes of illness, combats chronic inflammation, and cultivates authentic vitality, guiding individuals through a transformative process.

Tailored specifically to each individual, NoMAD journeys are meticulously crafted to rebalance the body, strengthen the mind, and rejuvenate overall health. By integrating ancestral practices with cutting-edge, innovative treatments—all under strict medical oversight—NoMAD Plans offer a personalised pathway to sustainable, long-lasting well-being that resonates with your unique life circumstances

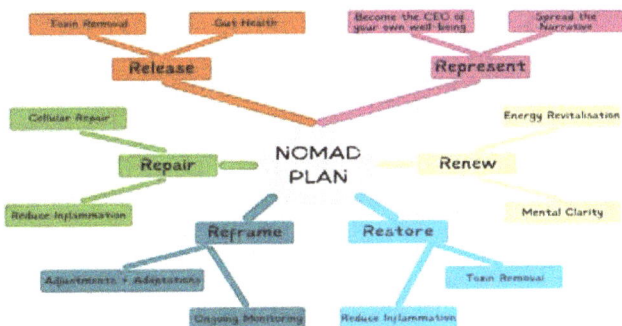

Levitas One:
"As Is In, As Is Out"

Reflecting the belief that our internal well-being is mirrored in our external environment. Founded by Dr. Ash Kapoor, Levitas One serves as the vehicle for delivering NoMAD's treatment plans. It envisions a healthcare future where patients are at the centre of a fully integrated, multidisciplinary approach. Guided by the 5 Rs—Restore, Release, Repair, Renew, and Revisit—Levitas One empowers self-care through personalised guidance and minimal intervention, promoting long-term health, balance, and sustainability.

Release   Represent

Repair   NoMad   Reframe

Renew   Restore

Contents

Forward: (need to find someone)

Preface

In the vast and intricate world of biological sciences, few molecules captivate the imagination as profoundly as peptides. These short chains of amino acids, often overshadowed by their larger protein cousins, hold within them the essence of life's complexity. Peptides are the unsung heroes, orchestrating a myriad of functions with precision and elegance—from signalling pathways that govern cellular communication to the very mechanisms that drive the immune response.

My own journey into the world of peptides began with a deep curiosity and a simple question: *how can such small molecules exert such powerful effects?* This question sparked a passion that has guided my research and exploration over the years. What I discovered was a world where peptides serve as nature's messengers, carrying out instructions with remarkable efficiency, influencing everything from metabolism to memory. Their roles are as diverse as they are essential, and their potential in therapeutic applications is vast and ever-expanding.

As I looked deeper into this field, I was struck not only by the scientific possibilities but also by the profound implications peptides have for the future of medicine. This realisation has driven my work and inspired this book, which is more than just a collection of knowledge; it is a testament to a journey of discovery, innovation, and a deep-seated belief in the power of these remarkable molecules.

In the pages that follow, you will find a synthesis of the latest research, practical insights, and visionary perspectives. But more importantly, you will find inspiration—drawn from the same wellspring that has fuelled my own journey. I hope to ignite in you the same passion to think beyond the conventional, to embrace the potential of peptides in ways that could revolutionise healthcare, biotechnology, and beyond.

This book is dedicated to the scientists, researchers, and innovators who, like me, have been captivated by the mystery and promise of peptides. It is also for the next generation of thinkers and doers who will carry this torch forward, exploring new frontiers with the courage and curiosity that define true discovery.

Let this book help navigate this journey together, into the world of peptides—where science and imagination converge, and where the future is waiting to be written.

Chapter 1
Introduction to Peptides in Contemporary Medicine

Analogy: The Peptides as Tiny Programmers

Imagine your body as a highly complex computer system. Just like any computer, it runs on specific programs that keep everything functioning smoothly. In this analogy, peptides are like tiny programmers, each with a specific task to ensure the system runs efficiently. Some of these programmers send important messages to different parts of the body, others manage repairs when something breaks down, and a few even handle security, protecting the body from harmful invaders. This chapter is about understanding these tiny programmers—peptides—and how they are becoming key players in modern medicine.

Understanding Peptides: The Basics

Peptides are tiny chains of amino acids, which are the building blocks of proteins—the essential substances that make up much of the structure and function of our bodies. While proteins are like large, complex machines, peptides are simpler and smaller, but they are critical. They typically consist of 2 to 50 amino acids linked together, forming a chain. The way these amino acids are arranged determines the shape and function of each peptide, much like how the arrangement of letters in a word determines its meaning.

In simpler terms, peptides act as messengers and regulators in the body. They help keep everything running smoothly, ensuring that our cells, tissues, and organs function properly. Peptides do this by interacting with cells, sending signals that can trigger specific actions, such as starting the repair process for damaged tissue or regulating how the body uses energy.

The Evolution of Peptide Therapy in Medicine

Over the years, our understanding of peptides has grown significantly. Traditional medicine often relies on drugs that affect large parts of the body at once, which can lead to side effects because they are not very specific. Peptides, on the other hand, can be designed to target specific areas or processes within the body, leading to more precise and effective treatments with fewer side effects.

Peptide therapy has shown potential in treating a wide range of health issues, from chronic diseases and ageing to hormonal imbalances and metabolic disorders. This chapter will introduce peptide therapy, exploring why these molecules are so important in modern medicine and how they are being used to improve health.

Historical Perspective: The Dawn of Peptide Research

The story of peptides in medicine began over a century ago, with a major breakthrough in 1921 when scientists Frederick Banting and Charles Best discovered insulin, a peptide hormone. Insulin quickly became a lifesaving treatment for diabetes, demonstrating the power of peptides in medicine.

In the early 20th century, interest in peptide hormones grew, leading to the discovery and development of several essential peptides. By the 1970s, scientists were able to create synthetic versions of peptides like vasopressin and oxytocin, which regulate things like water balance and social bonding.

In the late 20th and early 21st centuries, advancements in technology allowed scientists to study peptides in greater detail. Researchers began to see peptides not just as simple replacements for missing hormones but as tools that could be used to influence complex biological processes. This new understanding opened up possibilities for treating chronic diseases, managing the effects of ageing, and optimising overall health.

Basic Biology of Peptides: More Than Just Building Blocks

Although peptides are simpler and smaller than proteins, they play a wide variety of roles in the body. The specific job a peptide does depends on the order of its amino acids, which determines its shape and how it interacts with other molecules.

The Structure of Peptides

The structure of a peptide is like the blueprint for its function. The sequence of amino acids in a peptide determines how it folds and what shape it takes. This shape is crucial because it affects how the peptide interacts with cells and other molecules. For example, a simple, straight-chain peptide might be responsible for sending a quick signal. In contrast, a more complex, folded peptide might trigger a series of reactions in the body.

Roles of Peptides in the Body

- **Cell Signalling**: Peptides are like messengers, carrying important information between cells. For instance, insulin, a well-known peptide, helps regulate blood sugar levels by signalling cells to absorb glucose from the bloodstream.

- **Regulation of Gene Expression**: Some peptides can even influence which genes are turned on or off in our cells, leading to the production of specific proteins needed for various bodily functions. This process can promote growth, repair, or even the programmed death of damaged cells.

- **Immune Modulation**: Peptides also play vital roles in our immune system. Specific peptides can directly attack harmful bacteria or viruses, while others signal immune cells to ramp up or calm down their activity, helping the body fight infections or reduce inflammation.

- **Tissue Regeneration and Repair**: Peptides are crucial for healing and repairing tissues. Some peptides encourage the growth of new cells or blood vessels, which is essential when the body is repairing wounds or regenerating tissue after injury.

The Importance of Peptides in Modern Medicine

The introduction of peptide therapy has brought a new level of precision to medicine. Traditional drugs often affect broad areas of the body, which can lead to unwanted side effects. Peptides, however, can be specifically designed to interact only with specific cells or receptors, achieving the desired therapeutic effect with minimal unintended consequences.

Advantages of Peptide Therapy

- **High Specificity**: Peptides are highly specific, meaning they can be designed to target only certain cells or receptors in the body. This reduces the risk of side effects compared to traditional drugs.

- **Biocompatibility**: Since peptides are made from naturally occurring amino acids, the body easily recognises and breaks them down, reducing the likelihood of adverse reactions.

- **Versatility**: Peptides can be used to treat a wide range of conditions, from managing blood sugar levels in diabetes to promoting skin health and even enhancing cognitive function.

- **Personalisation**: Peptide therapy can be customised to fit the unique needs of each patient, allowing for more personalised and effective treatments.

Real-Life Applications of Peptide Therapy

Peptide therapy is already making a difference in many areas of medicine. As research continues, the ways we can use peptides are becoming more diverse and sophisticated. Here are some of the most significant applications:

1. Management of Diabetes

- **Insulin Therapy**: Insulin, one of the earliest and most well-known peptides, remains a crucial treatment for diabetes. Modern versions of insulin have been refined to better mimic the body's natural insulin, improving blood sugar control and reducing the risk of low blood sugar episodes.

- **GLP-1 Analogs**: These peptides help manage type 2 diabetes by enhancing the body's natural insulin response, reducing appetite, and promoting weight loss.

2. Anti-ageing and Skin Rejuvenation

- **GHK-Cu (Copper Peptide)**: This peptide helps maintain skin health by promoting the production of collagen, which keeps skin firm and elastic. It also has anti-inflammatory properties that speed up wound healing.

- **Matrixyl and Argireline**: These peptides are commonly found in anti-Ageing skincare products. They help reduce wrinkles by promoting collagen production and relaxing facial muscles, similar to how anti-wrinkle injectables works.

3. Cognitive Enhancement and Neuroprotection

- **Dihexa**: This peptide has shown promise in improving memory and learning, especially in conditions like Alzheimer's disease, by promoting the growth of new connections between brain cells.

- **Semax and Cerebrolysin**: These peptides are used to protect brain cells and enhance cognitive function. They are being explored as treatments for brain injuries and neurodegenerative diseases.

4. Immune System Modulation

- **Thymosin Alpha-1 (Tα1)**: This peptide boosts the immune system, helping the body fight off chronic infections and improving the effectiveness of cancer treatments.

- **LL-37**: An antimicrobial peptide that not only kills bacteria and viruses but also helps modulate the immune response, reducing inflammation and promoting healing.

- **Melanotan II**: Originally developed as a tanning agent, this peptide also influences the immune system and is being researched for its potential to treat skin conditions and enhance immune surveillance.

5. Tissue Regeneration and Healing

- **BPC-157**: Known for its ability to speed up the healing of injuries, BPC-157 promotes the formation of new blood vessels and collagen, which are crucial for tissue repair.

- **Thymosin Beta-4 (TB-500) and Epidermal Growth Factor (EGF)**: These peptides play essential roles in wound healing and tissue regeneration, making them valuable in treating injuries, burns, and chronic wounds.

6. Hormone Regulation and Replacement

- **Growth Hormone-Releasing Hormones (GHRH) and Growth Hormone-Releasing Peptides (GHRP)**: These peptides stimulate the natural release of growth hormone, which can help improve muscle mass, energy levels, and overall vitality.

- **Kisspeptin and Thyrotropin-Releasing Hormone (TRH)**: These peptides regulate reproductive and thyroid hormones, offering potential treatments for infertility and thyroid-related disorders.

Expanding the Horizons of Peptide Therapy

As we learn more about peptides, the possibilities for their use in medicine continue to grow. Researchers are discovering new peptides and new ways to use them, which could lead to treatments for a broader range of conditions, including cardiovascular diseases, obesity, and rare genetic disorders. The ability to create personalised peptide therapies tailored to individual needs is fascinating, as it represents a move toward more effective and safer treatments.

Key Takeaways

1. **Peptides: Crucial Bioregulators** - Peptides are essential molecules that help regulate many vital processes in the body.

2. **Precision Medicine with Peptides** - Peptide therapy offers a targeted, specific approach to treating various medical conditions with fewer side effects.

3. **Versatility of Peptide Applications** - Peptides are used in therapies ranging from diabetes management to skincare and cognitive enhancement.

4. **Revolutionary Shift in Medicine** - The development of peptide therapy marks a significant advancement in personalised and effective healthcare.

Recommendations for Further Exploration

1. **Clinical Research and Trials**: Ongoing research is needed to better understand the long-term effects and optimal dosages of peptide therapies. Expanded trials will help improve the safety and effectiveness of these treatments.

2. **Personalised Peptide Therapy**: With advancements in genetics and molecular biology, there is potential to create highly personalised peptide therapies that are tailored to each individual's unique genetic makeup and health status.

3. **Regulatory Frameworks**: As peptide therapies become more common, clear regulations are needed to ensure the safety, quality, and consistency of these treatments.

4. **Public Awareness and Education**: Educating both healthcare providers and the public about peptide therapy is essential for its broader acceptance and proper use.

5. **Interdisciplinary Collaboration**: The future of peptide therapy lies in collaboration between different scientific disciplines, including biology, pharmacology, and clinical medicine. This collaboration will drive innovation and uncover new applications for peptides.

By focusing on these areas, the medical community can unlock the full potential of peptide therapy, leading to more effective, personalised, and innovative treatments that improve patient outcomes across a wide range of health conditions.

Summary: Introduction to Peptides in Contemporary Medicine

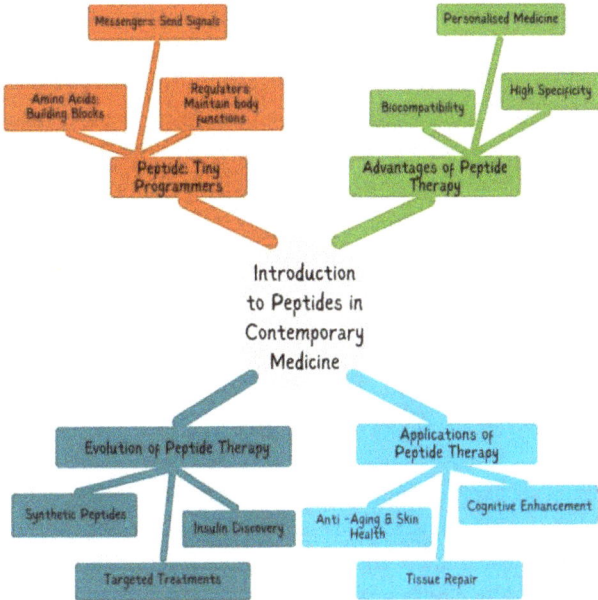

Chapter 2
Overview of Peptide Therapy

Imagine your body as a highly complex and finely tuned orchestra, with each instrument representing different cells, tissues, and organs. The conductor, who ensures everything works in harmony, is the signalling system that communicates instructions to these instruments. Peptide therapy can be thought of as a special set of musical notes or signals that the conductor uses to fine-tune the orchestra, ensuring that each part performs its role perfectly. These signals are tiny chains of amino acids, called peptides, that help regulate and optimise how the body functions. Just as the right musical note can bring an instrument into harmony, the right peptide can help a part of the body work better, whether it is to speed up healing, boost energy, or balance hormones.

How Peptides Work in the Body

Peptides are like messengers that deliver specific instructions to cells in the body. These instructions tell the cells how to act, much like a conductor telling musicians when to play louder or softer. When a peptide binds to a cell's surface, it triggers a series of reactions inside the cell that lead to a specific outcome, such as reducing inflammation, promoting tissue repair, or balancing hormones.

Signalling Through Receptors

One of the main ways peptides communicate with cells is by attaching special receptors on the cell's surface. Think of these receptors as locks, and peptides as keys. When the correct key (peptide) fits into the lock (receptor), it unlocks a series of actions inside the cell. Many peptides work through receptors known as GPCRs (G-protein-coupled receptors), which are involved in many of the body's essential functions, like managing stress, regulating mood, and controlling inflammation.

An Example of Peptide Action

Let us take the peptide BPC-157 as an example. BPC-157 is known for its ability to speed up healing. When BPC-157 attaches to the proper receptor, it triggers a chain reaction inside the cell that leads to faster tissue repair and reduced inflammation, much like how the conductor's cue can make the orchestra play faster or slower, depending on what the music requires.

Peptides and Gene Expression

In addition to working through receptors, some peptides can influence the way our genes are expressed. Genes are like instruction manuals for making everything in the body, and peptides can help turn certain pages in these manuals, either speeding up or slowing down the production of specific proteins. For example, a peptide might encourage the production of a protein that helps protect our cells from ageing, much like how a conductor might encourage a musician to play a specific part of a piece more prominently.

Telomere Protection

Epitalon is a peptide known for its ability to protect telomeres, the protective caps at the ends of our chromosomes. Telomeres shorten as we age, which is like the wear and tear on the pages of an instruction manual. Epitalon helps keep these telomeres longer, delaying the ageing process and keeping cells healthier for longer.

Pathways Inside the Cell

The instructions that peptides send often follow certain pathways inside the cell, leading to very specific results. These pathways are like roads that lead to different destinations, such as cell growth, repair, or immune response.

- **MAPK/ERK Pathway:**

This pathway is like a road that leads to tissue repair and growth, making it essential for healing injuries.

- **PI3K/AKT Pathway:**

This pathway is crucial for regulating metabolism and ensuring cells get the energy they need to survive.

- **JAK/STAT Pathway:**

This pathway is essential for controlling inflammation and immune responses, which are vital in fighting off infections and managing autoimmune diseases.

By using the correct peptide, we can guide the cell down the appropriate pathway, just like choosing the right road to reach a desired destination.

Personalised Peptide Therapy

One of the most exciting aspects of peptide therapy is how it can be personalised. Just as every musician in an orchestra might need different instructions based on their instrument, each person's body has unique needs. Peptide therapy can be tailored to these individual needs, making it a highly effective treatment option.

Using Biomarkers to Guide Treatment

Personalisation often starts with checking biomarkers, which are indicators in your blood or tissue that show how your body is functioning. Think of biomarkers as the tuning notes a conductor listens to before starting a performance. By understanding your biomarkers, a healthcare provider can choose the suitable peptides to help bring your body into better balance.

Examples of Customised Therapy

- **Growth Hormone Deficiency:**

For someone with low levels of growth hormone, a peptide like Sermorelin can be used to naturally boost the body's own growth hormone production rather than using synthetic hormones.

- **Chronic Inflammation:**

For conditions like chronic inflammation, biomarkers can help determine whether a peptide like Thymosin Alpha-1 might help reduce inflammation and strengthen the immune system.

- **Metabolic Imbalances:**

If you have issues with metabolism, such as insulin resistance, a peptide like MOTS-c can be used to improve insulin sensitivity and help regulate blood sugar levels.

Ongoing Monitoring and Adjustment

Peptide therapy is not a one-size-fits-all solution. It requires ongoing monitoring, just like a conductor might adjust the orchestra's performance based on how the music is flowing. Healthcare providers will regularly check your progress and make adjustments to your peptide regimen as needed, ensuring the therapy continues to meet your specific needs.

Case Example

Consider someone recovering from surgery. Initially, they might use BPC-157* to speed up tissue repair. As they heal, their doctor might switch them to a different peptide that focuses more on long-term recovery and prevention of scar tissue.

Personalised Dosing and Delivery

The dose and delivery method of peptides can also be customised. Some people might need a higher or lower dose depending on how their body processes the peptide. Peptides can be delivered in different ways too—some might be taken as a pill, while others might be injected under the skin for faster results.

Addressing Common Misconceptions

Despite the many benefits of peptide therapy, some misconceptions can make people hesitant to try it.

Peptides Are Not Steroids

One common myth is that peptide therapy is similar to using anabolic steroids, which are often associated with bodybuilding and harmful side effects. However, peptides and steroids are very different. While steroids can cause a wide range of unwanted side effects, peptides are much more targeted in their actions, focusing on specific areas of the body without causing widespread hormonal changes.

Safety and Regulation

Another misconception is that peptides are unregulated or unsafe. In reality, therapeutic peptides undergo rigorous testing to ensure they are both safe and effective. In the United States, for example, the FDA closely monitors and approves peptide therapies, ensuring that they meet strict standards before being allowed on the market.

Real-Life Examples

Peptides like Semaglutide (used for managing diabetes) have been approved by the FDA after extensive testing. This approval process involves numerous clinical trials to confirm their safety and effectiveness.

Key Health Issues Addressed by Peptide Therapy

Peptide therapy can be used to address a wide range of health concerns. Here are some key areas where peptides can make a significant difference:

Mitochondrial Health

Mitochondria are the energy factories of our cells. When they don't work correctly, it can lead to fatigue, muscle weakness, and other issues. Peptides like MOTS-c and SS-31 help keep these energy factories running smoothly, improving overall energy levels and cellular health.

MOTS-c: Enhancing Energy

MOTS-c is a peptide that helps regulate how your body uses energy, especially in conditions like type 2 diabetes, where energy regulation is impaired. By improving how your body processes sugar, MOTS-c can help manage blood sugar levels and support weight loss.

SS-31: Protecting Cells

SS-31 is another peptide that protects the mitochondria from damage, ensuring they continue to produce energy efficiently. This is particularly important in conditions like heart failure, where energy production in cells is critical.

Pineal Gland Health

The pineal gland produces melatonin, which regulates our sleep-wake cycles. Peptides like Endoluten can help restore proper function to this gland, improving sleep quality and overall well-being.

Endoluten: Regulating Sleep

Endoluten helps regulate melatonin production, particularly in older individuals whose natural melatonin levels might have decreased. By restoring normal sleep patterns, Endoluten can improve mood, energy levels, and even immune function.

Stress and Cognitive Function

The hypothalamus plays a key role in regulating stress and cognitive function. Peptides like Selank and Cerebrolysin can help manage stress and improve mental clarity by modulating how the hypothalamus functions.

Selank: Reducing Anxiety

Selank is a peptide that reduces anxiety and improves mental focus, making it an excellent option for those dealing with chronic stress or anxiety disorders.

Cerebrolysin: Supporting Brain Health

Cerebrolysin is a peptide blend that supports cognitive function, particularly after a stroke or in conditions like Alzheimer's disease. It helps protect brain cells and encourages the growth of new connections between neurons.

Liver Health

The liver is essential for detoxification and metabolism. Peptides like Ovagen support liver health by promoting the repair and regeneration of liver cells.

Ovagen: Supporting Liver Function

Ovagen helps the liver recover from damage, whether it is due to fatty liver disease, hepatitis, or other conditions. By promoting liver cell regeneration, Ovagen can help improve overall metabolic health.

Gut Health

The gastrointestinal (GI) tract is another area where peptide therapy can be highly effective. Peptides like BPC-157 help heal the gut lining, reducing inflammation and promoting overall digestive health.

BPC-157: Healing the Gut*

BPC-157* is known for its ability to heal the gut lining, making it an excellent option for those with conditions like Crohn's disease or irritable bowel syndrome (IBS). By reducing inflammation and promoting tissue repair, BPC-157* can significantly improve gut health.

Musculoskeletal Health

Peptide therapy is also beneficial for treating injuries to muscles, tendons, and joints. Peptides like TB-500 and GHK-Cu promote tissue repair and reduce inflammation, speeding up recovery from injuries.

TB-500: Promoting Recovery

TB-500 is a peptide that helps repair muscle and tendon injuries, making it particularly useful for athletes or those recovering from surgery.

GHK-Cu: Enhancing Skin and Joint Health

GHK-Cu is a peptide that supports skin and joint health by promoting collagen production and reducing inflammation. It is often used to improve the appearance of the skin and to treat conditions like osteoarthritis.

The Importance of Patient Involvement

Successful peptide therapy relies on the active involvement of patients. Understanding the treatment plan, having realistic expectations, and sticking to the prescribed therapy are all essential for achieving the best results.

Educating Patients

Healthcare providers should take the time to explain how peptide therapy works, what to expect, and how to manage any potential side effects. This helps patients stay informed and engaged with their treatment.

Understanding the Plan

Patients should know why specific peptides have been chosen and how they will help. This knowledge encourages patients to be more committed to their therapy.

Managing Expectations

Patients need to understand that peptide therapy often takes time to show results. Setting realistic expectations helps maintain patient satisfaction and ensures they stick with the treatment.

Regular Monitoring

Ongoing monitoring allows healthcare providers to adjust the therapy as needed, ensuring it remains effective. Patients should be encouraged to regularly check in with their provider and report any changes in their symptoms or overall health.

Building a Strong Therapeutic Relationship

A successful outcome in peptide therapy also depends on a solid relationship between the patient and the healthcare provider. Open communication, trust, and collaboration are crucial to ensuring that the treatment is effective and that the patient feels supported throughout the process.

Summary: Overview of Peptide Therapy

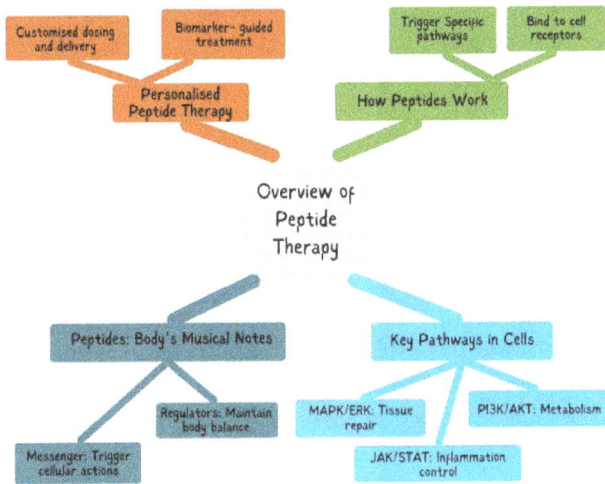

Chapter 3
Understanding Peptides

This chapter, "Understanding Peptides," serves as a comprehensive yet accessible introduction to the fascinating world of peptides, designed to cater to both medical professionals and those without a medical background who are interested in learning about this important area of biology and medicine.

The chapter begins with a relatable analogy that compares the human body to a high-tech factory, where peptides act as specialised workers responsible for a wide range of critical tasks. This analogy sets the stage for a deeper exploration of peptides, helping readers grasp the significance of these molecules in maintaining the body's health and function.

The Building Blocks: What Are Peptides?

The first section of the chapter delves into the fundamental structure of peptides, explaining that they are chains of amino acids linked together in specific sequences. The chapter breaks down the classification of peptides into three categories: oligopeptides, polypeptides, and proteins, based on the length of their amino acid chains. By using simple analogies, such as comparing amino acids to beads on a string, the chapter makes it easy for readers to visualise how the arrangement of these "beads" determines the peptide's function.

The chapter also emphasisess the importance of the three-dimensional shape of peptides, explaining how this shape allows them to interact with other molecules in the body. For example, some peptides might twist into spirals or fold into sheets, which enables them to fit precisely into specific binding sites on target molecules. This concept is crucial for understanding how peptides perform their diverse roles in the body.

What Do Peptides Do?

In this section, the chapter explores the wide range of functions that peptides perform in the body. It presents peptides as multi-talented workers who take on various roles, from acting as messengers that transmit signals between different parts of the body, to serving as defenders that protect against infections.

The chapter breaks down the critical roles of peptides into easily understandable categories:

- **Messengers**: Peptides as hormones or neurotransmitters that send vital signals across the body, with insulin being highlighted as an example of a peptide hormone that regulates blood sugar levels.

- **Immune Helpers**: The role of peptides in guiding and modulating the immune response, helping the body fight off infections and control inflammation.

- **Defenders**: An explanation of antimicrobial peptides that act as the body's security guards, destroying harmful pathogens.

- **Repair and Growth**: Peptides involved in healing and regeneration, such as growth factors that promote tissue repair.

- **Enzymes and Regulators**: Peptides that speed up or regulate biochemical reactions, ensuring the body functions smoothly.

This section is designed to provide readers with a clear understanding of how essential peptides are to the body's everyday operations, maintaining health and responding to challenges like injury or disease.

What Makes Peptides Special?

Peptides hold immense promise as therapeutic agents due to their unique characteristics. It emphasises that peptides are naturally compatible with the human body, which reduces the risk of adverse reactions when they are used as treatments. This section breaks down the key advantages of peptides:

Natural Fit: Peptides and the Human Body

The body's inherent familiarity with peptides contributes to their safety and effectiveness as therapies. Because peptides are already part of the body's biological systems, they are less likely to cause unwanted immune responses, making them ideal for personalised medicine.

Staying Power: Keeping Peptides Active

A key challenge in developing peptide-based therapies lies in their stability within the body. It explains how peptides can be quickly broken down by enzymes and outlines the innovative strategies researchers use to enhance their durability. These strategies include chemical modifications, the creation of peptide mimetics (peptide look-alikes that are more resistant to degradation), and protective delivery systems like nanoparticles.

Precision Targeting: Hitting the Right Spot

One of the most powerful attributes of peptides is their remarkable ability to target specific tissues or cells within the body. The chapter explains how this precision reduces the risk of side effects, as peptides can be designed to interact only with specific receptors on certain cells, such as targeting cancer cells while sparing healthy tissue. This targeted action is contrasted with traditional drugs, which often affect multiple systems in the body, leading to unintended consequences.

Supporting Rather Than Replacing

Unlike traditional hormone therapies that simply replace what the body lacks, peptides take a more nuanced approach. This approach helps maintain the body's natural balance, reducing the risk of dependency and making peptides a safer option for long-term use.

Safety First: Peptides and Your Health

Reassurance comes from the fact that peptides, with their natural origins in the body, are typically well-tolerated and pose a lower risk of toxicity compared to synthetic drugs. Furthermore, they break down into harmless amino acids after fulfilling their therapeutic role, minimising lingering side effects. This inherent precision further enhances their safety profile by reducing the likelihood of unintended interactions within the body.

Peptides: The Future of Medicine

Looking ahead, ongoing research continues to uncover new applications for peptides in treating a wide range of health conditions. Their unique properties—such as compatibility, stability, targeted action, and safety—make them versatile tools in modern medicine.

This exploration of peptide therapies also touches on the potential for personalised medicine, where treatments can be designed to meet the specific needs of individual patients. Imagine new delivery systems, such as nanoparticles and implantable devices, enhancing the effectiveness of these therapies by ensuring they reach their intended targets with pinpoint accuracy.

Ultimately, peptides represent a promising frontier in medicine, with the potential to revolutionise how we treat and prevent diseases. Their significance in advancing healthcare and improving patient outcomes is undeniable, paving the way for exciting developments in peptide-based medicine.

Summary: Understanding Peptides

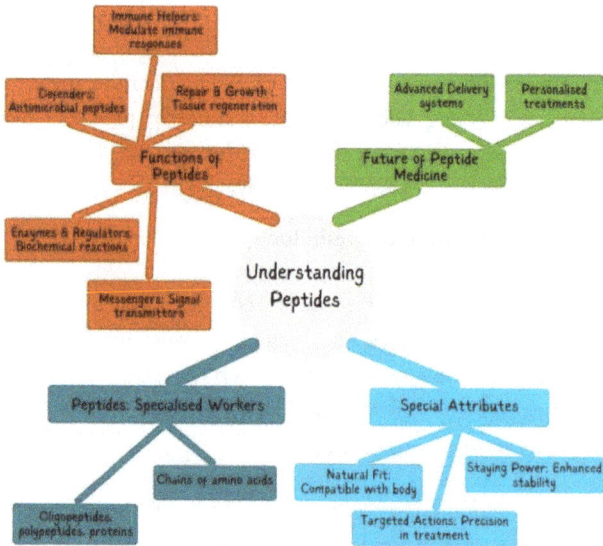

Immune Helpers:
Modulate immune
responses

Defenders:
Antimicrobial peptides

Repair & Growth :
Tissue regeneration

Functions of
Peptides

Enzymes & Regulators:
Biochemical reactions

Messengers: Signal
transmitters

Advanced Delivery
systems

Personalised
treatments

Future of Peptide
Medicine

Understanding
Peptides

Peptides: Specialised Workers

Chains of amino acids

Oligopeptides,
polypeptides, proteins

Special Attributes

Natural Fit:
Compatible with body

Staying Power: Enhanced
stability

Targeted Actions: Precision
in treatment

Chapter 4
Understanding the Classification of Peptide Bioregulators: Cytomaxes and Cytogens

Imagine your body as a well-tuned orchestra, where each instrument (organ) plays its part in harmony. Just like a conductor who ensures that every musician is playing correctly, your body has natural regulators that keep everything running smoothly. However, with time, some instruments might fall out of tune or get worn out, affecting the overall performance. Peptide bioregulators, the focus of this chapter, act like expert technicians or assistant conductors, stepping in to fine-tune the instruments, ensuring that every part of the orchestra (your body) performs at its best.

We will explore two main types of peptide bioregulators: Cytomaxes and Cytogens. These peptides are unique molecules that help your body's organs and tissues function optimally, whether by supporting long-term health or providing a quick fix when something goes wrong. This information is useful not only for doctors but also for anyone interested in maintaining or improving their health.

Cytomaxes: Natural Helpers for Long-Term Health

Cytomaxes are like the natural repair crew in your body. They are derived from natural sources, specifically from animal organs and tissues. They are designed to help keep your body's systems running smoothly over the long term. Think of them as a gentle tune-up for your body's machinery, ensuring that everything continues to work well as you age.

How Cytomaxes Work

Cytomaxes work by interacting with specific cells in your body, much like how a key fits into a lock. Once they connect with these cells, they trigger processes that help the cells function better, repair themselves, and stay healthy. This is especially important as we get older because our cells can become less efficient over time. By supporting these natural processes, Cytomaxes help keep your organs and tissues in good shape, reducing the risk of age-related problems.

Key Cytomax Products and Their Benefits

Here is a closer look at some of the key Cytomax products and how they can help different parts of your body:

1. **BONOMARLOT®: Supporting Blood Health**
   o **What It Does**: BONOMARLOT® helps maintain the health of your blood, particularly important for producing new blood cells.

   o **When It is Useful**: This product is beneficial if you are dealing with anaemia (a condition where your body does not have enough healthy red blood cells) or undergoing cancer treatment, which can affect blood production.

   o **How It Works**: It supports the parts of your body responsible for making blood, helping them work more effectively and produce the cells your body needs.

2. **BONOTHYRK®: Keeping Your Bones Strong**
   o **What It Does**: BONOTHYRK® focuses on the health of your parathyroid glands, which are crucial for managing calcium levels and bone strength.

   o **When It is Useful**: This product is particularly helpful in preventing osteoporosis, a condition where bones become weak and brittle.

o   **How It Works**: By helping your parathyroid glands function better, BONOTHYRK® ensures that your body maintains the right balance of calcium, which is essential for strong bones.

3. **VENTFORT®: Maintaining Healthy Blood Vessels**
   o   **What It Does**: VENTFORT® supports the health of your blood vessels, ensuring they remain flexible and strong.

   o   **When It is Useful**: This is particularly useful if you have conditions like high blood pressure or atherosclerosis (hardening of the arteries).

   o   **How It Works**: VENTFORT® helps keep the walls of your blood vessels in good shape, reducing the risk of complications related to poor circulation.

4. **VISOLUTEN®: Protecting Your Eyes**
   o   **What It Does**: VISOLUTEN® is designed to support the health of your eyes, particularly the delicate tissues that are crucial for vision.

   o   **When It is Useful**: This product can be beneficial if you are dealing with eye conditions like glaucoma or cataracts, which can impair your vision.

   o   **How It Works**: VISOLUTEN® promotes the repair and renewal of eye tissues, helping to preserve your eyesight as you age.

5. **VLADONIX®: Boosting Your Immune System**
   o   **What It Does**: VLADONIX® helps strengthen your immune system, which is your body's defence against infections and diseases.

   o   **When It is Useful**: This product is particularly beneficial if your immune system is weakened, or if you are looking to prevent conditions like cancer.

o   **How It Works**: VLADONIX® enhances the function of your immune cells, making your body better equipped to fight off harmful invaders.

6.   **GLANDOKORT®: Managing Stress and Hormones**
o   **What It Does**: GLANDOKORT® supports the health of your adrenal glands, which are responsible for managing stress and balancing hormones.

o   **When It is Useful**: This product is helpful if you are dealing with stress-related disorders or hormonal imbalances.

o   **How It Works**: By supporting your adrenal glands, GLANDOKORT® helps your body cope with stress more effectively and maintains hormonal balance.

7.   **GOTRATIX®: Enhancing Muscle Health**
o   **What It Does**: GOTRATIX® is designed to support muscle health and function, making it ideal for those who are physically active or experiencing muscle fatigue.

o   **When It is Useful**: This product is beneficial for athletes or anyone dealing with muscle weakness.

o   **How It Works**: GOTRATIX® helps your muscles recover and stay strong, improving endurance and reducing fatigue.

8.   **ZHENOLUTEN®: Supporting Women's Reproductive Health**
o   **What It Does**: ZHENOLUTEN® focuses on the health of the ovaries, which are crucial for women's reproductive health.

o   **When It is Useful**: This product is beneficial if you are looking to regulate your menstrual cycle or enhance fertility.

o **How It Works**: ZHENOLUTEN® supports the normal function of ovarian cells, helping to maintain hormonal balance and improve reproductive outcomes.

9. **LIBIDON®: Promoting Prostate Health**

o **What It Does**: LIBIDON® is designed to support prostate health, addressing common issues like chronic prostatitis (inflammation of the prostate) and erectile dysfunction.

o **When It is Useful**: This product is beneficial for men experiencing prostate-related issues.

o **How It Works**: LIBIDON® helps reduce inflammation in the prostate and improves urinary and sexual function.

10. **PIELOTAX®: Enhancing Kidney Health**

o **What It Does**: PIELOTAX® supports the health of your kidneys, which are essential for filtering waste from your body.

o **When It is Useful**: This product is beneficial for those with kidney problems, such as nephropathy or renal insufficiency.

o **How It Works**: PIELOTAX® helps regenerate kidney cells, improving their ability to filter and eliminate waste.

11. **SIGUMIR®: Supporting Joint and Bone Health**

o **What It Does**: SIGUMIR® focuses on the health of your cartilage and bones, which are essential for movement and physical stability.

o **When It is Useful**: This product is helpful if you are dealing with conditions like arthritis or osteochondrosis (a condition affecting the bones and cartilage).

o **How It Works**: SIGUMIR® promotes the repair of cartilage tissue, improving joint mobility and reducing pain.

12. **STAMAKORT®: Protecting Your Stomach**
    o **What It Does**: STAMAKORT® supports the health of your stomach lining, which is vital for proper digestion.

    o **When It is Useful**: This product is beneficial if you are experiencing stomach issues like gastritis or ulcers.

    o **How It Works**: STAMAKORT® helps regenerate the cells of the stomach lining, protecting against damage and improving digestion.

13. **SUPREFORT®: Enhancing Pancreatic Function**
    o **What It Does**: SUPREFORT® supports the pancreas, an organ that plays a crucial role in digestion and blood sugar regulation.

    o **When It is Useful**: This product is helpful in managing conditions like diabetes or digestive issues.

    o **How It Works**: SUPREFORT® helps the pancreas produce the necessary enzymes and hormones, improving digestion and glucose metabolism.

14. **SVETINORM®: Supporting Liver Health**
    o **What It Does**: SVETINORM® focuses on the liver, which is essential for detoxifying the body and managing metabolism.

    o **When It is Useful**: This product is particularly beneficial for individuals with liver disease or those exposed to toxins.

    o **How It Works**: SVETINORM® promotes liver cell regeneration, helping the liver function more effectively in detoxifying the body.

15. TAXOREST®: Promoting Respiratory Health

o **What It Does**: TAXOREST® supports the health of your bronchial tubes, which are vital for breathing.

o **When It is Useful**: This product is useful for those with chronic respiratory conditions like bronchitis or COPD.

o **How It Works**: TAXOREST® helps regenerate the cells in your bronchial tubes, improving respiratory function and reducing symptoms.

16. TESTOLUTEN®: Supporting Male Reproductive Health

o **What It Does**: TESTOLUTEN® focuses on supporting male reproductive health, particularly in enhancing sperm motility and libido.

o **When It is Useful**: This product is beneficial for men experiencing infertility or reduced sexual function.

o **How It Works**: TESTOLUTEN® promotes the regeneration of testicular cells, improving sperm production and enhancing sexual performance.

17. THYREOGEN®: Enhancing Thyroid Function

o **What It Does**: THYREOGEN® supports the thyroid gland, which is crucial for regulating metabolism and energy levels.

o **When It is Useful**: This product is helpful in managing thyroid disorders like hypothyroidism or hyperthyroidism.

o **How It Works**: THYREOGEN® helps the thyroid gland produce hormones that regulate metabolism, improving overall energy and metabolic balance.

18. **CERLUTEN®: Supporting Brain and Nervous System Health**
    o **What It Does**: CERLUTEN® focuses on the health of your brain and nervous system, which are essential for cognitive function and mental well-being.

    o **When It is Useful**: This product is beneficial for conditions like Alzheimer's disease, depression, or other neurological disorders.

    o **How It Works**: CERLUTEN® promotes the regeneration of nerve cells, improving cognitive function and reducing symptoms of neurological diseases.

19. **CHELOHART®: Promoting Heart Health**
    o **What It Does**: CHELOHART® supports the health of your heart muscle, which is crucial for pumping blood throughout your body.

    o **When It is Useful**: This product is particularly beneficial for individuals with heart disease, such as ischemic heart disease.

    o **How It Works**: CHELOHART® promotes the regeneration of heart muscle cells, improving heart function and reducing the risk of heart disease.

20. **CHITOMUR®: Enhancing Bladder Health**
    o **What It Does**: CHITOMUR® supports the health of your bladder, which is essential for storing and eliminating urine from the body.

    o **When It is Useful**: This product is beneficial for those with bladder issues like chronic cystitis or urinary incontinence.

    o **How It Works**: CHITOMUR® helps regenerate bladder cells, improving urinary control and reducing symptoms of bladder disorders.

21. **ENDOLUTEN®: Supporting Neuroendocrine Health**
   o **What It Does**: ENDOLUTEN® focuses on the neuroendocrine system, which regulates hormones and affects sleep patterns.

   o **When It is Useful**: This product is beneficial for individuals experiencing hormonal imbalances, sleep disorders, or chronic fatigue.

   o **How It Works**: ENDOLUTEN® promotes the regeneration of neuroendocrine cells, improving hormone regulation and enhancing sleep quality.

Cytogens: Synthetic Helpers for Rapid Response

While Cytomaxes are like your body's long-term maintenance crew, Cytogens are more like the emergency responders. These peptides are synthesised in the lab, designed to act quickly when something in your body needs immediate attention. They are especially useful for addressing acute conditions where fast intervention is required in order to restore normal function.

How Cytogens Work

Cytogens are made to imitate your body's natural peptides, which are short chains of amino acids that play a crucial role in regulating various bodily functions. When you introduce Cytogens into your body, they bind to specific receptors on your cells, much like a key fitting into a lock. This triggers a series of actions inside the cell that lead to repair and regeneration. Because they are designed for quick action, Cytogens are particularly effective in situations where your body needs a rapid response.

Key Cytogen Products and Their Benefits

Here are some of the essential Cytogen products and how they can help your body recover quickly:

1. **VESUGEN®: Supporting Blood Vessel Health**
   o **What It Does**: VESUGEN® is designed to support the health of your blood vessels, ensuring they function properly.

   o **When It is Useful**: This product is particularly beneficial for conditions like varicose veins or atherosclerosis, where blood flow is compromised.

   o **How It Works**: VESUGEN® helps your blood vessels stay flexible and strong, reducing the risk of complications related to poor circulation.

2. **CRYSTAGEN®: Boosting Immune Function**
   o **What It Does**: CRYSTAGEN® is designed to give your immune system a boost, helping your body fight off illness.

   o **When It is Useful**: This product is especially useful when you are recovering from an illness or want to prevent future sickness.

   o **How It Works**: CRYSTAGEN® enhances the activity of your immune cells, making your body better equipped to defend against infections and diseases.

3. **CARTALAX®: Supporting Joint and Cartilage Health**
   o **What It Does**: CARTALAX® focuses on maintaining healthy joints and cartilage, which are crucial for movement and flexibility.

   o **When It is Useful**: This product is beneficial for conditions like arthritis or osteoporosis, where joint health is compromised.

   o **How It Works**: CARTALAX® promotes the regeneration of cartilage, improving joint function and reducing pain and stiffness.

4. **OVAGEN®: Enhancing Liver Function**
   o **What It Does**: OVAGEN® is designed to support liver health, which is essential for detoxification and overall metabolism.

   o **When It is Useful**: This product is particularly useful for those with liver conditions or who need to support their liver's detoxifying functions.

   o **How It Works**: OVAGEN® promotes the regeneration of liver cells, helping the liver process toxins more effectively and improving metabolic health.

5. **PINEALON®: Supporting Brain Health**
   o **What It Does**: PINEALON® is designed to promote the health of your brain, improving cognitive function and mental clarity.

   o **When It is Useful**: This product is especially beneficial for individuals dealing with cognitive decline or mental health issues like depression.

   o **How It Works**: PINEALON® supports the regeneration of brain cells, enhancing memory, focus, and overall brain function.

6. **CHONLUTEN®: Enhancing Respiratory and Digestive Health**
   o **What It Does**: CHONLUTEN® supports both your respiratory and digestive systems, which are closely connected in maintaining overall health.

   o **When It is Useful**: This product is useful for those dealing with chronic bronchitis, asthma, or digestive issues.

   o **How It Works**: CHONLUTEN® helps regenerate cells in your respiratory and digestive tracts, improving function and reducing symptoms associated with chronic conditions.

Conclusion

In-Depth Exploration of Peptide Bioregulators

This comprehensive analysis has focused on the two main categories of peptide bioregulators: Cytomaxes and Cytogens.

Distinct Categories with Unique Therapeutic Benefits

- **Cytomaxes**: Derived from natural sources, these peptides primarily support the long-term health of organs and tissues. They play a crucial role in promoting cellular renewal and maintaining physiological balance, which is essential for preventing age-related diseases.

- **Cytogens**: As synthetic peptides, Cytogens are designed to provide a rapid therapeutic response, particularly effective in treating acute conditions. Their ability to activate cellular regeneration makes them invaluable in scenarios where immediate intervention is required.

Importance of Understanding Applications and Mechanisms

A deep understanding of the specific applications, mechanisms of action, and therapeutic potential of each peptide product is critical. This knowledge is essential for healthcare providers and patients alike, enabling them to choose the most effective, targeted therapies for a wide range of health conditions.

LeverAgeing Peptides for Improved Health Outcomes

By harnessing the unique properties of Cytomaxes and Cytogens, significant improvements in health outcomes can be achieved. These peptides are particularly beneficial in the prevention and management of age-related diseases and chronic conditions, offering a promising avenue for enhancing overall health and well-being.

Summary: Understanding the Classification of Peptide Bioregulators
(a)

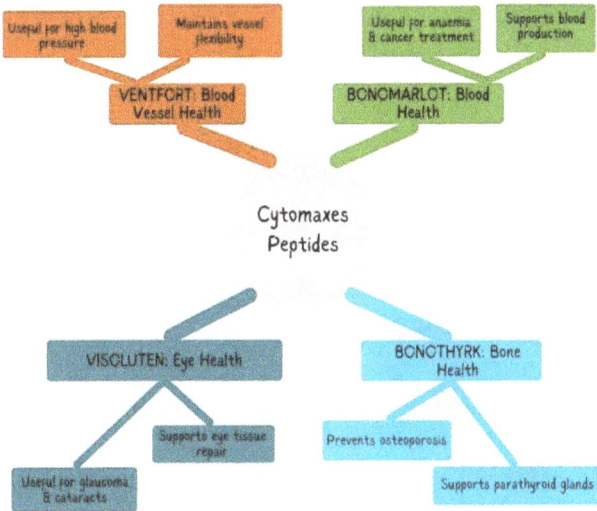

Useful for high blood pressure

Maintains vessel flexibility

VENTFORT: Blood Vessel Health

Useful for anaemia & cancer treatment

Supports blood production

BONOMARLOT: Blood Health

Cytomaxes Peptides

VISOLUTEN: Eye Health

Supports eye tissue repair

Useful for glaucoma & cataracts

BONOTHYRK: Bone Health

Prevents osteoporosis

Supports parathyroid glands

(b)

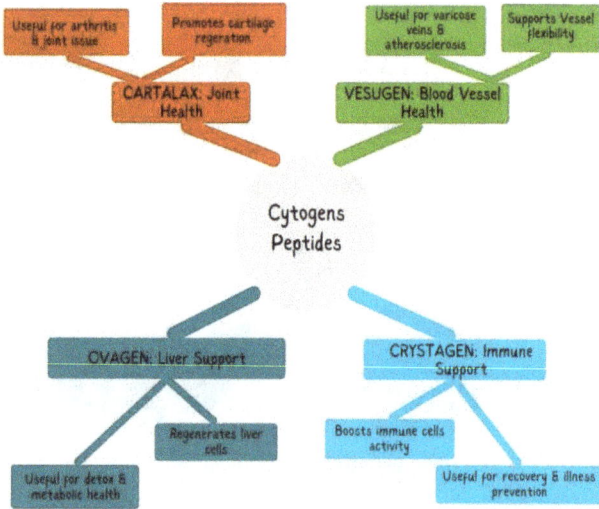

Useful for arthritis & joint issue

Promotes cartilage regeration

CARTALAX: Joint Health

Useful for varicose veins & atherosclerosis

Supports Vessel flexibility

VESUGEN: Blood Vessel Health

Cytogens Peptides

OVAGEN: Liver Support

Regenerates liver cells

Useful for detox & metabolic health

CRYSTAGEN: Immune Support

Boosts immune cells activity

Useful for recovery & illness prevention

(c)

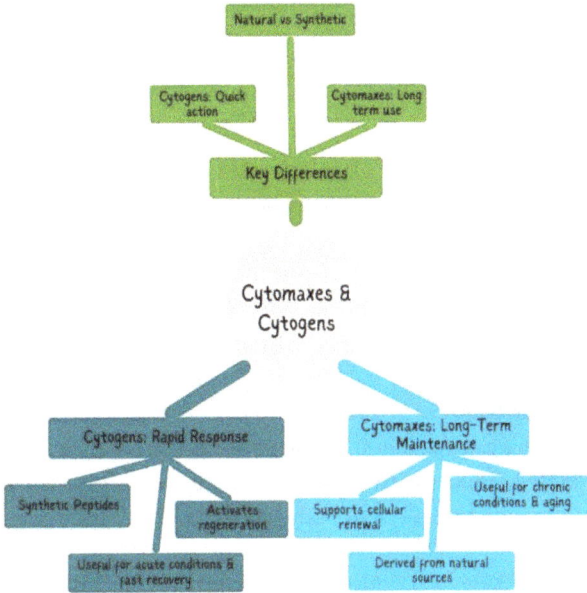

Chapter 5: Revilab ML Series: Multifunctional Peptide Solutions

Introduction: Understanding the Revilab ML Series

Imagine your body is like a large, complex orchestra, with each organ and system acting as a different section—strings, woodwinds, brass, and percussion. Just like how each section must play in harmony to produce beautiful music, every part of your body needs to function together in balance to maintain good health. Now, what if a few sections of the orchestra start to fall out of tune? Perhaps the strings are slightly off-pitch, or the brass section is playing too loudly. The Revilab ML Series is like a highly skilled conductor, using specialised tools (in this case, peptides) to bring each section back into harmony. These tools are designed to address various issues simultaneously, helping the orchestra—your body—play beautifully once again.

This chapter will explore how the Revilab ML Series, a line of innovative, multifunctional peptide-based solutions, works to restore and maintain this balance in your body. Whether you are a healthcare provider looking for new therapeutic options or someone simply interested in improving your well-being, this chapter will guide you through the science, benefits, and practical use of these products in a way that is easy to understand.

The Science Behind the Revilab ML Series

Peptides, the primary components of the Revilab ML Series, are small chains of amino acids—the building blocks of proteins. Think of them as the "notes" that make up the "music" of life in your body. These peptides play a critical role in regulating various biological processes, much like how musical notes can form different melodies depending on their arrangement. By using specific peptides, the

Revilab ML products can send precise signals to your body's cells, guiding them to perform certain functions, such as repairing tissues, reducing inflammation, or boosting immunity.

The concept behind the Revilab ML Series is based on biomimicry, which means these peptides mimic the natural processes your body already uses. For example, suppose your immune system is under stress. In that case, specific peptides in these products can signal your body to strengthen its defences, much like a conductor signalling the orchestra to play louder when needed.

Each product in the Revilab ML Series is carefully formulated to include a unique combination of these peptides, along with other beneficial ingredients like antioxidants (which help protect your cells from damage), vitamins, and minerals. These additional ingredients work together with the peptides to create a more robust, more effective solution, similar to how different instruments in an orchestra combine to create a richer sound.

One of the standout features of the Revilab ML Series is its method of delivery: sublingual administration. This means the products are taken by placing them under the tongue, where they are quickly absorbed into the bloodstream. This method bypasses the digestive system, allowing the active ingredients to enter your body faster and work more effectively. It is like giving the conductor a direct line to the orchestra, ensuring that instructions are communicated clearly and promptly.

Key Revilab ML Products: Simple Explanations for Powerful Solutions

Now, let us dive into the specific products within the Revilab ML Series, each designed to address particular health concerns. Whether you are dealing with ageing, heart health, or cognitive function, there is likely a product tailored to meet your needs.

Revilab ML 01: Anti-Ageing and Anticancer Complex

Revilab ML 01 is like a dual-purpose tool designed to help slow down the ageing process and provide some protection against cancer. It contains peptides that help your body's cells regenerate and repair themselves, which is crucial as you age. Additionally, it includes ingredients that boost your immune system and protects your cells from oxidative stress—a kind of damage that contributes to both ageing and the development of cancer. Think of it as a high-quality tune-up for your body, helping to keep everything running smoothly and preventing major issues before they start.

Revilab ML 02: Anti-Anemic Complex

This product is designed for people who struggle with anaemia. In this condition, your body does not have enough healthy red blood cells to carry oxygen to your tissues. Revilab ML 02 acts like a specialised repair kit, stimulating the production of these red blood cells. It is beneficial for individuals undergoing treatments like chemotherapy, which can cause anaemia. By improving blood health, this product helps restore energy and vitality, much like ensuring that every section of the orchestra has enough breath to play their instruments effectively.

Revilab ML 03: Nervous System and Retina Support

If you are concerned about maintaining sharp thinking and clear vision as you age, Revilab ML 03 is designed with you in mind. This product contains peptides that protect the nerve cells in your brain and the cells in your retina (the part of your eye that helps you see clearly). It is like a protective shield for your nervous system and eyes, helping to keep these crucial parts of your body functioning well, much like ensuring that the conductor's sheet music is free of smudges and easy to read.

Revilab ML 04: Cardiovascular Health Complex

Heart health is the focus of Revilab ML 04, a product that targets the cardiovascular system—the heart and blood vessels. The peptides in this formulation help improve the health of your blood vessels, regulate blood pressure, and reduce inflammation, all of which are critical for preventing heart disease. Think of it as an ongoing maintenance service for your heart, ensuring that the "rhythm section" of your body (your heart and blood vessels) keeps the beat steady and strong.

Revilab ML 05: Respiratory System Complex

For those with chronic respiratory conditions like asthma or COPD, Revilab ML 05 is designed to enhance lung function and support easier breathing. The peptides in this product help reduce inflammation in the lungs and improve oxygen exchange, much like fine-tuning the wind instruments in an orchestra to ensure clear, strong notes. This product is particularly valuable for anyone living in areas with high pollution or for those with a history of breathing difficulties.

Revilab ML 06: Gastrointestinal Health Support

Digestive issues can be uncomfortable and disruptive, but Revilab ML 06 is here to help. This product supports the health of your gastrointestinal tract, improving digestion and protecting the lining of your stomach and intestines. It is like giving your digestive system a gentle yet effective cleaning, ensuring everything flows smoothly. This is particularly beneficial for individuals with conditions like irritable bowel syndrome (IBS) or ulcers.

Revilab ML 07: Men's Health Complex

Revilab ML 07 is tailored for men's health, with a focus on prostate health and enhancing libido. The peptides in this product help maintain the health of the prostate, reducing the risk of conditions like benign prostatic hyperplasia (BPH), which is common as men age. Additionally, this formula includes ingredients known to enhance sexual function, making it a comprehensive tool for men who want to maintain their vitality. It is like fine-tuning the bass section of the orchestra, ensuring that the foundation remains solid and steady.

Revilab ML 08: Women's Health Complex

For women, Revilab ML 08 supports ovarian function and helps regulate hormonal balance, which is crucial for managing issues like menstrual irregularities, menopause symptoms, or overall reproductive health. The peptides in this product work much like a hormonal conductor, ensuring that everything stays in harmony, whether you are dealing with monthly cycles or life changes like menopause.

Revilab ML 09: Joint Health Support

As we age, joint pain and stiffness can become common, but Revilab ML 09 is designed to help. This product supports joint health by promoting the regeneration of cartilage and improving bone density. It is like adding lubrication to the moving parts of an instrument, ensuring smooth, pain-free movement. This is especially helpful for people with conditions like arthritis or osteoporosis.

How to Use the Revilab ML Series: Practical Tips for Maximum Benefits

Using the Revilab ML products is simple and designed to fit easily into your daily routine. These products are taken sublingually, meaning you place a few drops under your tongue and hold them there for about one to two minutes before swallowing. This method allows the active ingredients to bypass your digestive system and enter your bloodstream quickly, ensuring they start working faster. It is a bit like taking a shortcut in a race—getting to the finish line more rapidly and efficiently.

Each product in the Revilab ML Series is designed with specific goals in mind, but they can also be combined to address multiple health concerns simultaneously. For example, suppose you are worried about both heart health and cognitive function. In that case, you might use Revilab ML 04 and Revilab ML 03 together. This approach allows you to create a customised health plan that fits your unique needs, much like how a conductor might arrange a piece of music to highlight different sections of the orchestra at various times.

Integrating Revilab ML Series into Treatment Protocols

Whether you are a healthcare provider or someone looking to take charge of your health, the Revilab ML Series offers a versatile approach to wellness. These products can be used as part of a broader treatment plan or on their own, depending on your needs. It is essential to consider your overall health, any existing conditions, and other treatments or medications you might be using. The Revilab ML products are designed to complement other therapies, making them a valuable addition to any health regimen.

In summary, the Revilab ML Series provides a sophisticated yet accessible approach to health management. By harnessing the power of peptides and delivering them efficiently through sublingual administration, these products offer rapid and effective solutions for a wide range of health concerns. Whether you are looking to prevent future issues or manage existing conditions, the Revilab ML Series can help you achieve and maintain optimal health.

Summary: Revilab ML Series: Multifunctional Peptide Solutions

(a)

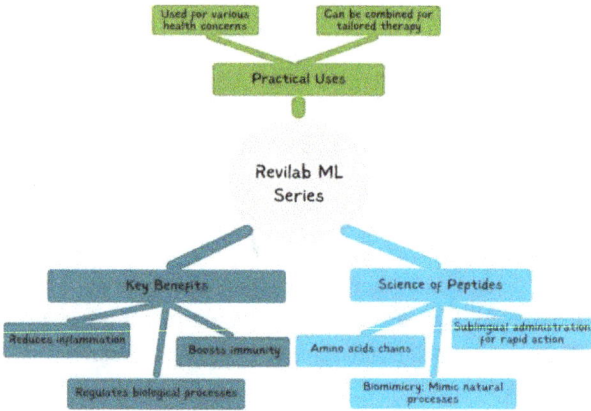

(b)

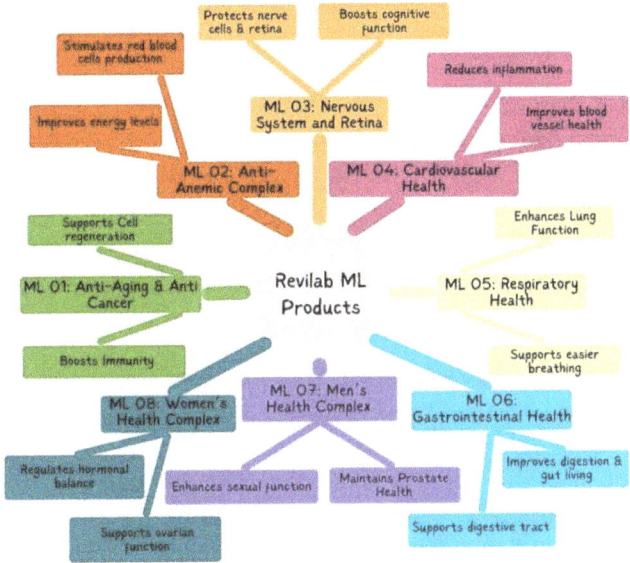

Chapter 6: Peptide Therapy Protocols: Step-by-Step Approaches

Introduction: Understanding Peptide Therapy

Imagine your body as a vast, intricate garden, where each organ and system represents different types of plants—flowers, shrubs, trees, and grasses. Just like in any garden, all these plants need the right conditions—water, sunlight, and nutrients—to thrive. But what happens when some plants begin to wither, weeds start to take over, or the soil becomes depleted? The garden loses its harmony, and the beauty of the landscape fades.

Peptide therapy is like having a master gardener at your disposal, equipped with specialised tools to restore balance and vitality to your garden. These tools—peptides—are carefully chosen to address the specific needs of each plant. Whether it is reviving a wilting flower (a fatigued organ), removing invasive weeds (toxins), or enriching the soil (supporting cellular function), peptide therapy works to bring your body's garden back to full bloom.

This is a comprehensive guide for healthcare professionals who wish to integrate peptide therapy into their clinical practice, as well as for individuals seeking to understand how these therapies can support their health. It offers detailed, evidence-based protocols for addressing various health concerns, including mitochondrial health, detoxification, anti-glycation strategies, and more. Each protocol is designed with precise steps for clinicians and includes user-friendly patient information leaflets to ensure clarity and adherence.

Protocols for Healthcare Professionals

These protocols provide a structured and systematic approach to peptide therapy, ensuring that treatments are tailored to meet the unique needs of each patient. Detailed guidance is provided on the selection, administration, and monitoring of therapeutic peptides and related interventions.

1. Mitochondrial Health and Energy Support

Objective:

To optimise mitochondrial function, which is crucial for sustaining energy production, enhancing metabolic efficiency, and maintaining overall cellular health. This protocol is particularly beneficial for patients who have chronic fatigue syndrome, metabolic dysfunctions, or age-related declines in energy.

Think of mitochondria as the power plants of your garden, providing the energy needed for all the plants to grow and thrive. If these power plants are underperforming, the entire garden suffers. This protocol is designed to revitalise these power plants, ensuring that your garden remains lush and vibrant.

Components:

- **Glycine:**
  An amino acid that acts as a precursor to glutathione, a critical antioxidant that helps protect your mitochondria from damage and supports their efficient function. Imagine glycine as a nutrient-rich fertiliser that enhances the soil's ability to nourish your plants.

- **N-Acetylcysteine (NAC):**
  NAC is a precursor to glutathione, playing a vital role in detoxification and protecting mitochondria from oxidative stress. It acts like a protective barrier, shielding the power plants in your garden from harsh environmental elements.

- **L-Carnitine:**
  Essential for the transport of fatty acids into mitochondria,

where they are oxidised to produce energy. Think of L-Carnitine as the transport trucks delivering fuel to your power plants, ensuring they have what they need to generate energy.

- **Magnesium:**
  A cofactor required for hundreds of enzymatic reactions, including those involved in energy metabolism and mitochondrial function. Magnesium is like the irrigation system in your garden, ensuring that water (in this case, energy) flows smoothly to all the plants.

- **Coenzyme Q_{10}:**
  Integral to the electron transport chain, CoQ_{10} is crucial for the production of ATP, the energy currency of the cell. It is like the spark that ignites the engines in your power plants, driving energy production.

Peptides:

- **SS-31 (Elamipretide):**
  A mitochondrial-targeted peptide that stabilises the inner mitochondrial membrane reducing oxidative stress and enhancing ATP production. SS-31 acts like a specialised gardener tending directly to the roots, ensuring they remain strong and resilient.

- **Humanin:**
  A peptide that improves mitochondrial function and protects against oxidative stress and cellular apoptosis. Humanin is like a vigilant garden caretaker, constantly monitoring and repairing damage to keep the plants (cells) healthy.

Administration:

- **Glycine and NAC:**
  Administered orally, typically on a daily basis, much like regularly adding compost to your garden to maintain soil health.

- **L-Carnitine and Magnesium:**
  Can be given orally or intravenously, depending on patient-specific needs and the desired speed of effect. This is akin to using drip irrigation for a slow, steady supply of water or a sprinkler system for a more immediate effect.

- **Coenzyme Q10:**
  Best absorbed when taken with meals containing fats; administered orally. This ensures that the "fuel" is delivered effectively to the power plants.

- **Peptides:**
  Administered subcutaneously, with dosage and frequency customised based on the patient's mitochondrial function and energy levels. This is like applying a targeted fertiliser directly to the roots that need it most.

Monitoring:

Regular laboratory assessments, including ATP levels, oxidative stress markers (e.g., 8-OHdG, F2-isoprostanes), and patient-reported outcomes related to energy levels, cognitive function, and overall well-being. Just as a gardener regularly checks the health of the soil and plants, continuous monitoring ensures that the treatment is effective and adjustments can be made as needed.

2. Detoxification Protocol

Objective:

To enhance the body's detoxification pathways, particularly for patients exposed to high levels of environmental toxins, those with impaired liver function, or those undergoing treatments like chemotherapy that increase toxic burden.

Imagine the detoxification process as weeding your garden. Weeds (toxins) can choke out healthy plants, robbing them of nutrients and space. This protocol is designed to efficiently remove these weeds, allowing your garden to flourish.

Components:

- **Glutathione:**
  A master antioxidant and detoxifying agent that plays a central role in liver detoxification processes and protection against oxidative damage. Glutathione is like a powerful weed killer, selectively targeting and removing harmful elements from your garden.

- **Alpha-Lipoic Acid:**
  A potent antioxidant that also chelates heavy metals and recycles other antioxidants, including glutathione. It acts like a multi-purpose gardening tool, capable of both removing weeds and enriching the soil.

- **Milk Thistle Extract:**
  Contains silymarin, which supports liver detoxification, regeneration, and protection against toxin-induced damage. Milk thistle is like a natural soil conditioner that helps your garden recover after a harsh winter or a bout of pests.

Peptides:

- **Liver-Regenerative Peptides (e.g., LR-3 IGF-1):**
  Promote liver cell regeneration and enhance detoxification capacity. These peptides are like a gardener's secret blend of nutrients that revitalise tired soil, helping it bounce back stronger than ever.

- **Toxin-Binding Peptides:**
  Specialised peptides designed to bind and neutralise specific toxins, facilitating their removal from the body. Think of these peptides as a specialised weeding tool that targets only the unwanted plants, leaving the rest of your garden untouched.

Administration:

- **Glutathione:**
  Administered intravenously for immediate effect, with oral or sublingual forms for maintenance.

- **Alpha-Lipoic Acid and Milk Thistle:**
  Administered orally, with dosing adjusted based on the level of toxic exposure and liver function tests.

- **Peptides:**
  Administered subcutaneously, with dosage tailored according to the specific toxins involved and the patient's detoxification capacity.

Monitoring:

Frequent monitoring of liver function tests (e.g., ALT, AST, GGT), oxidative stress markers, and toxin levels (e.g., heavy metals, environmental pollutants) to ensure effective detoxification and adjust treatment as needed. Regular checks, like inspecting the garden for new weeds or signs of stress, help maintain the overall health of the garden.

3. Primary Peptide Therapy

Objective:

To employ primary peptides for the treatment of specific health conditions, including hormonal imbalances, degenerative diseases, and organ-specific dysfunctions, aiming for targeted and personalised therapeutic outcomes.

Consider primary peptide therapy as the tailored care for each unique plant in your garden. Just as different plants have different needs—some require more sunlight, others need more water—this therapy addresses the specific needs of each organ or system, ensuring that each one thrives in its optimal conditions.

Key Peptides:

- **Endoluten:**
  Enhances pineal gland function, crucial for regulating circadian rhythms and maintaining hormonal balance. Endoluten is like adjusting the sunlight exposure in your garden, ensuring that each plant gets the right amount of light to bloom at the right time.

- **Ovagen:**
  Supports liver health and function, aiding in detoxification and overall metabolic regulation. Ovagen is like enhancing the soil quality in a particular area of your garden, ensuring that plants in that zone have the nutrients they need to grow strong.

- **Vesugen:**
  Promotes vascular health by maintaining the integrity of blood vessels and reducing inflammation, which is pivotal in preventing cardiovascular diseases. Vesugen acts like strengthening the garden's irrigation system, ensuring that water (blood) flows freely to all parts of the garden.

Administration:

- **Endoluten:**
  Administered subcutaneously, typically in the evening to coincide with the body's natural melatonin production cycle.

- **Ovagen and Vesugen:**
  Administered subcutaneously, with treatment schedules and dosages personalised to address the specific organ dysfunctions or health goals of the patient.

Monitoring:
Comprehensive monitoring, including hormonal profiles, liver enzymes, vascular health markers (e.g., CRP, homocysteine), and patient-reported symptoms to track therapeutic efficacy and make necessary adjustments to the treatment plan.

4. Anti-Glycation Strategy

Objective:

To mitigate glycation, a process that leads to the formation of harmful compounds known as advanced glycation end-products (AGEs), which contribute to ageing and chronic diseases by damaging proteins, lipids, and DNA.

Imagine glycation as a slow-acting toxin that gradually turns your vibrant garden into a brittle, dry landscape. This strategy is designed to neutralise that toxin, preserving the freshness and vitality of your garden.

Components:

- **Revilab Anti-AGE:**
  A peptide-based formulation designed to reduce the accumulation of AGEs in the body and protect against their harmful effects. Revilab Anti-AGE is like a specialised treatment that rejuvenates withering plants, restoring their health and vigor.

- **Carnosine:**
  An anti-glycation agent that inhibits the formation of AGEs and provides antioxidant protection against oxidative stress. Carnosine is like a protective coating for your plants, preventing the damaging effects of harsh weather or pests.

- **Alpha-Lipoic Acid:**
  Reduces oxidative stress and enhances the body's defences against glycation, further supported by its ability to regenerate other antioxidants. Alpha-Lipoic Acid acts like a nutrient-rich mulch, insulating the soil and protecting the roots from damage.

Peptides:

- **Additional Anti-Glycation Peptides:**
  These peptides target specific tissues or organs most affected by glycation, such as the kidneys, eyes, or skin.

Administration:

- **Revilab Anti-AGE and Carnosine:**
  Administered orally, with dosage adjusted based on the patient's age, glycation levels, and overall health status. These supplements are like a regular feeding schedule for your garden, ensuring that nutrients are delivered consistently to keep plants strong and healthy.

- **Peptides:**
  Administered subcutaneously, with a treatment schedule designed to match the severity of glycation and specific health conditions.

Monitoring:

Regular assessments of AGE levels (e.g., carboxymethyllysine, pentosidine), oxidative stress biomarkers, and clinical indicators of ageing or disease progression, such as skin elasticity, renal function, or visual acuity.

Patient Information Leaflets

To complement the protocols provided for healthcare professionals, the following patient information leaflets are designed to offer clear, accessible explanations of the treatments. These leaflets aim to ensure that patients understand the purpose, process, and expected outcomes of their therapy, promoting better adherence and more successful results.

1. Mitochondrial Health and Energy Support

Purpose:

This treatment aims to enhance your energy levels by supporting your cells' ability to produce energy. It is instrumental if you are feeling fatigued or if your doctor has identified issues with your mitochondrial function.

What to Expect:

Patients often report feeling more energetic and mentally sharp within a few weeks. Over time, you may also notice improvements in physical stamina and overall well-being.

Instructions:

- **Supplements:**
  Take your glycine, NAC, L-Carnitine, and magnesium as prescribed. It is important to follow the instructions carefully to get the full benefits.

- **Peptides:**
  These injections are designed to directly support your cellular energy production. Depending on your treatment plan, you might receive these in the clinic or learn how to administer them at home.

Monitoring:

We will regularly check in on your progress, including how you are feeling day-to-day. Periodic blood tests may be done to make sure the treatment is working as expected.

2. Detoxification Protocol

Purpose:

This protocol is designed to help your body get rid of harmful toxins and improve your liver's health, which is key to overall well-being.

What to Expect:

As your body detoxifies, you might notice clearer skin, better digestion, and more energy. This is a sign that the treatment is working.

Instructions:

- **Supplements:**
  Your treatment includes glutathione, alpha-lipoic acid, and milk thistle. These supplements are essential for protecting and cleansing your liver.

- **Peptides:**
  The peptides you will receive help your liver detoxify and regenerate. These injections may be given in the clinic, or you may be taught how to administer them at home.

Monitoring:

Regular check-ups and blood tests will be scheduled to monitor your liver function and the progress of your detoxification process.

3. Primary Peptide Therapy

Purpose:

This treatment involves using specific peptides to target and improve your health in areas like hormone balance, liver function, or vascular health.

What to Expect:

The results can vary, but many patients start to feel better within a few weeks to months, depending on the condition being treated. The goal is to address the root causes of your health issues.

Instructions:

- **Peptides:**
  You will receive injections of specific peptides based on your individual health needs. It is crucial to stick to the prescribed schedule to get the best results.

Monitoring:

Your health will be closely monitored, and adjustments to your treatment will be made as necessary. This might include blood tests and other evaluations to measure progress.

4. Anti-Glycation Strategy

Purpose:

This treatment is designed to protect your body from the damaging effects of glycation, which can speed up ageing and contribute to chronic diseases.

What to Expect:

As the treatment takes effect, you might notice improvements in your skin, energy levels, and overall health.

Instructions:

- **Supplements:**
  You will be taking Revilab Anti-AGE and carnosine to help prevent and repair damage caused by glycation.

- **Peptides:**
  These injections are part of your treatment plan to protect your body from glycation. Your healthcare provider will give you instructions on how to administer these, if necessary.

Monitoring:

We will monitor your progress with regular check-ups and possibly blood tests to make sure the treatment is effective.

This enhanced and detailed narrative provides both healthcare professionals and patients with a clear understanding of the protocols, their purposes, and the expected outcomes, ensuring that peptide therapy can be effectively integrated into clinical practice and personal health management.

Summary: Peptide Therapy Protocols: Step-by-Step Approaches
(a)

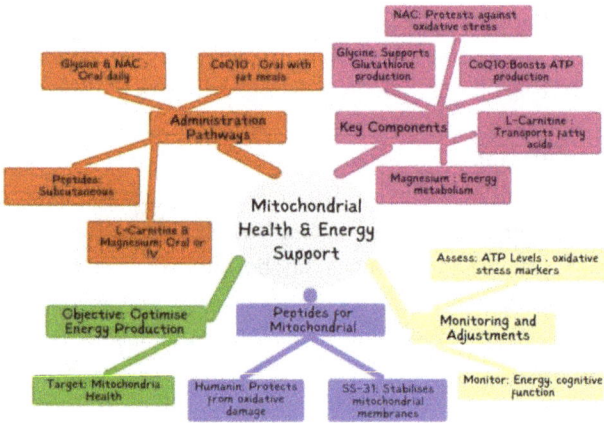

(b)

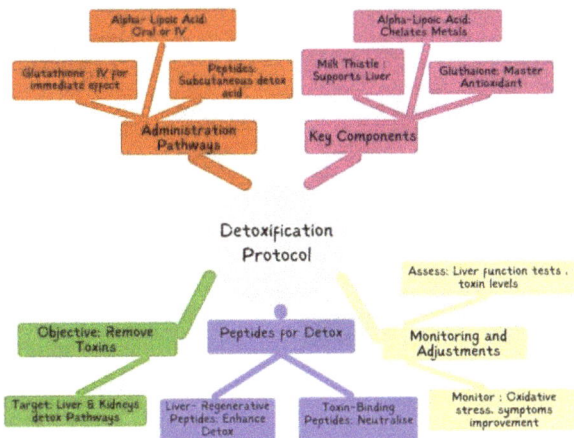

Detoxification Protocol

Administration Pathways
- Alpha-Lipoic Acid: Oral or IV
- Glutathione : IV for immediate effect
- Peptides: Subcutaneous detox acid

Key Components
- Alpha-Lipoic Acid: Chelates Metals
- Milk Thistle : Supports Liver
- Gluthaione: Master Antioxidant

Objective: Remove Toxins
- Target: Liver & Kidneys detox Pathways

Peptides for Detox
- Liver- Regenerative Peptides: Enhance Detox
- Toxin-Binding Peptides: Neutralise

Monitoring and Adjustments
- Assess: Liver function tests , toxin levels
- Monitor : Oxidative stress, symptoms improvement

(c)

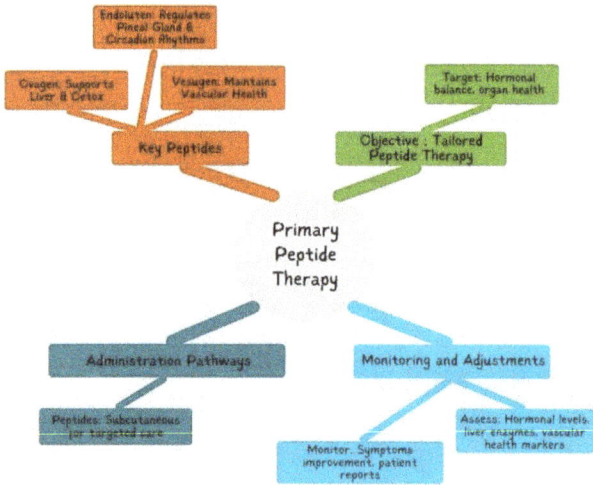

Primary Peptide Therapy

Key Peptides
- Endoluten: Regulates Pineal Gland & Circadian Rhythms
- Ovagen: Supports Liver & Detox
- Vesugen: Maintains Vascular Health

Objective : Tailored Peptide Therapy
- Target: Hormonal balance, organ health

Administration Pathways
- Peptides: Subcutaneous for targeted care

Monitoring and Adjustments
- Monitor: Symptoms improvement, patient reports
- Assess: Hormonal levels, liver enzymes, vascular health markers

(d)

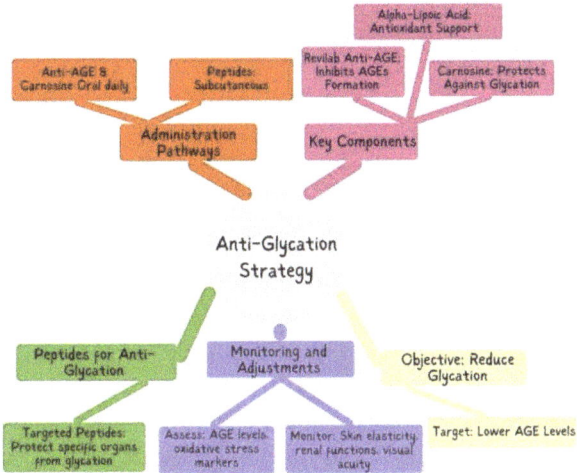

Alpha-Lipoic Acid: Antioxidant Support

Anti-AGE S Carnosine Oral daily

Peptides: Subcutaneous

Revilab Anti-AGE: Inhibits AGEs Formation

Carnosine: Protects Against Glycation

Administration Pathways

Key Components

Anti-Glycation Strategy

Peptides for Anti-Glycation

Monitoring and Adjustments

Objective: Reduce Glycation

Targeted Peptides: Protect specific organs from glycation

Assess: AGE levels, oxidative stress markers

Monitor: Skin elasticity, renal functions, visual acuity

Target: Lower AGE Levels

Chapter 7: Advanced Applications of Peptide Therapy

Introduction

Imagine the human body as a complex orchestra, where each instrument represents a different organ, and the music they produce together is your overall health. When one instrument falls out of tune—whether it is due to chronic exposure to toxins, the slow decline of the brain in neurodegenerative diseases, or the inevitable wear and tear of ageing—the harmony of your body is disrupted. Peptide therapy acts like a skilled conductor, bringing each instrument back in tune, restoring the balance, and allowing your body to perform at its best. In this chapter, we explore the advanced applications of peptide therapy, a cutting-edge approach designed to address some of the most challenging health issues we face today.

Protocols for Healthcare Professionals and Patients

1. Chronic Intoxication Management

Understanding the Problem: Chronic intoxications occur when your body is exposed to harmful substances over a long period. These toxins can come from the environment, such as pollutants in the air, water, and food, or from lifestyle choices, like smoking or alcohol consumption. Over time, these toxins accumulate in your body, overwhelming your natural detoxification systems—the liver, kidneys, and immune system.

The Peptide Therapy Approach: Peptide therapy works by enhancing the body's natural ability to detoxify and protect itself from the damaging effects of these toxins. Think of it as hiring a specialised cleaning crew for your body's detoxification pathways, with each member of the crew (each peptide) having a specific role:

- **Detoxification Pathway Support:**

Peptides such as Glutathione and Methylation Support Peptides act as key detoxifiers, helping the liver break down and eliminate toxins more efficiently. These peptides are like heavy-duty cleaning agents that scrub away accumulated toxins, allowing your organs to function more effectively.

Supplements like N-acetylcysteine (NAC) and alpha-lipoic acid (ALA) provide additional support, much like giving your cleaning crew extra tools to get the job done faster and more thoroughly. They work by boosting your body's antioxidant levels, reducing oxidative stress, and aiding in the safe removal of toxins through the kidneys and liver.

- **Organ Protection:**

To protect vital organs like the liver and kidneys from the onslaught of toxins, peptides such as Thymosin Beta-4 and Hepatoprotective Peptides are used. These peptides can be compared to putting protective shields around your organs, helping to repair any damage that has already occurred and preventing further harm.

- **Heavy Metal Detoxification:**

For those with high levels of heavy metals in their system, specific chelating peptides act like magnets, binding to metals like mercury, lead, and arsenic and pulling them out of the body. These peptides are combined with mineral supplements to replenish any essential nutrients that might be lost during the detox process, ensuring your body stays balanced.

- **Immune System Support:**

Chronic exposure to toxins can weaken your immune system, leaving you more vulnerable to infections and illnesses. Immunomodulatory peptides, like Thymosin Alpha-1, work to strengthen your immune response, much like reinforcing the walls of a fortress to keep invaders out.

Implementation for Healthcare Professionals: Each patient's treatment plan is customised based on their specific toxic load, organ function, and overall health. Regular monitoring through blood tests, imaging, and symptom tracking is essential to ensure the treatment is working effectively and making adjustments as needed. For patients, understanding that this process may take time and requires consistent monitoring is crucial for successful outcomes.

2. Neurodegenerative Disease Support

Understanding the Problem: Neurodegenerative diseases, such as Alzheimer's, Parkinson's, and Multiple Sclerosis, are like a slow-moving storm that gradually erodes the brain's cognitive abilities, memory, and physical coordination. These conditions are characterised by the progressive degeneration of neurons, the cells responsible for transmitting information throughout the brain and body.

The Peptide Therapy Approach: Peptide therapy offers a way to slow down or even halt the progression of these diseases by protecting neurons, enhancing brain function, and improving blood flow to the brain. Think of it as fortifying a castle under siege, where each peptide plays a role in strengthening the defences and ensuring the brain remains as functional as possible for as long as possible.

- **Cognitive Function Enhancement:**

Peptides such as Cerebrolysin and Dihexa are like brain enhancers, boosting cognitive abilities by promoting the growth of new neurons and strengthening the connections between them. These peptides help sharpen memory, focus, and overall mental clarity, making it easier for patients to navigate daily life.

- **Neuroprotection:**

Peptides like BDNF (Brain-Derived Neurotrophic Factor) mimetics and Semax act as protectors of the brain, shielding neurons from damage caused by inflammation and oxidative stress. Think of these peptides as guardians standing watch over your brain cells, ensuring they stay healthy and functional.

- **Improved Blood Flow:**

Peptides that improve vascular health, such as Endothelin Receptor Antagonists, work to ensure the brain receives adequate blood supply, much like clearing blocked pathways to allow vital supplies to reach a besieged city. This increased blood flow ensures that the brain gets the oxygen and nutrients it needs to function optimally.

- **Anti-inflammatory Strategy:**

Chronic inflammation is a crucial contributor to the progression of neurodegenerative diseases. Peptides such as Melanocortin analogues are used to reduce inflammation in the brain, acting like cooling agents that calm the fires of inflammation, protecting neurons from further damage.

- **Mitochondrial Support:**

The mitochondria are the powerhouses of the cells, responsible for producing energy. In neurodegenerative diseases, mitochondrial function often declines. Peptides like MOTS-c and SS-31 work to restore mitochondrial function, much like ensuring the castle's power supply remains strong, providing the energy needed to maintain defences.

Implementation for Healthcare Professionals: This protocol requires a highly personalised approach, with regular cognitive assessments, brain imaging, and blood tests to monitor the progression of the disease and the effectiveness of the treatment. For patients, understanding that neurodegenerative diseases require long-term management and a commitment to ongoing treatment and monitoring is vital.

3. Anti-Ageing Strategy with Peptide Therapy

Understanding the Problem: Ageing is a natural process, but it can be compared to the slow decay of a once-majestic building. Over time, the effects of wear and tear—both internal and external—become more apparent, leading to wrinkles, decreased energy levels, and a general decline in vitality. While we cannot stop ageing, peptide therapy offers a way to repair and maintain the structure, slowing down the ageing process and preserving health and vitality.

The Peptide Therapy Approach: Peptide therapy in anti-ageing works by addressing the underlying causes of ageing, such as cellular damage, hormonal decline, and reduced collagen production. Think of it as a comprehensive renovation plan for your body, where peptides act as master craftsmen, repairing damage, reinforcing weak spots, and enhancing overall vitality.

- **Cellular Rejuvenation:**

Peptides like Epitalon and FOXO4-DRI are used to rejuvenate cells by extending telomere length (the protective caps on the ends of chromosomes) and clearing out senescent cells (old, damaged cells that no longer function properly). This is like restoring the foundation of a building, ensuring it remains strong and stable for years to come.

- **Skin Health Improvement:**

Peptides that boost collagen production, such as GHK-Cu and Pal-GHK, are used to reduce wrinkles, improve skin elasticity, and promote a youthful appearance. These peptides are like specialised masons who repair cracks in the walls and restore the building's exterior to its former glory.

- **Vitality and Energy Enhancement:**

Peptides that stimulate the production of human growth hormone (HGH), such as CJC-1295 and Ipamorelin, work to increase energy levels, improve muscle mass, and enhance physical endurance. This is akin to upgrading the building's infrastructure, ensuring it has the strength and vitality to function at its best.

- **Antioxidant Support:**

The ageing process is accelerated by oxidative stress, which can be compared to rust slowly eating away at the metal framework of a building. Antioxidant peptides and supplements, like Superoxide Dismutase (SOD) mimetics and Glutathione boosters, help to neutralise free radicals and protect cells from oxidative damage, keeping the structure intact.

- **Lifestyle Modifications:**

In addition to peptide therapy, lifestyle factors such as diet, exercise, and stress management play a critical role in the anti-ageing strategy. Adopting a nutrient-rich diet, engaging in regular physical activity, and practising stress-reduction techniques are like ensuring the building is well-maintained, preventing further decay and preserving its condition.

Implementation for Healthcare Professionals: The anti-ageing protocol is tailored to each individual's unique genetic makeup, lifestyle, and health status. Regular monitoring of biomarkers related to ageing, such as telomere length, hormone levels, and skin health, allows for adjustments to the treatment plan as needed. For patients, understanding that ageing is a continuous process and that peptide therapy is a tool to slow down rather than stop the process is essential.

Patient Information Leaflets

Purpose and Importance: For patients undergoing peptide therapy, understanding the purpose and process of their treatment is critical to achieving the best outcomes. The patient information leaflets provided in this chapter are designed to bridge the gap between complex medical concepts and practical, easy-to-understand guidance. Whether you are a healthcare professional or a patient, these leaflets serve as a roadmap for the journey through peptide therapy.

Content Overview:

- **Introduction to Peptide Therapy:**

A clear explanation of what peptide therapy is, how it works within the body, and the specific benefits it offers for your condition. This section aims to demystify peptide therapy, providing you with a solid understanding of the treatment you are receiving.

- **Specific Treatment Information:**

Each leaflet is customised to the specific protocol you are following—whether it is for detoxification, neurodegenerative disease support, or anti-ageing. It explains the role of each peptide in your treatment, how it contributes to your overall health, and what results you can expect over time.

- **Administration Guidelines:**

Detailed instructions on how to administer your peptides correctly, including dosage, timing, and any special precautions you need to take. This section is like a user manual for your therapy, ensuring you use the treatments safely and effectively.

- **Expected Effects and Timeline:**

Information on the timeline of your treatment, including when you might start seeing results and what to expect during the course of therapy. It also covers potential side effects and how to manage them so you know what's normal and when to seek medical advice.

- **Monitoring and Follow-Up:**

Guidance on the importance of regular follow-up appointments, including blood tests and other assessments, to track your progress. This section emphasises the collaborative nature of your treatment, where ongoing communication between you and your healthcare provider is vital to success.

- **Lifestyle and Supportive Measures:**

Recommendations for lifestyle changes that complement your peptide therapy, such as dietary tips, exercise routines, and stress management strategies. These suggestions are like additional tools that help you get the most out of your treatment, supporting your overall health and well-being.

- **Frequently Asked Questions (FAQs):**

A section addressing common questions and concerns, providing straightforward answers and additional resources for those who want to learn more. This part of the leaflet is designed to put your mind at ease, ensuring you feel informed and confident in your treatment journey.

Accessibility: These leaflets are designed with clarity in mind, using simple language and visuals to make complex information easy to understand. Available in both printed and digital formats, they are easily accessible, ensuring that you have the information you need at your fingertips whenever you need it.

Conclusion

Chapter 7 serves as an invaluable resource for both healthcare professionals and patients, providing advanced protocols for the application of peptide therapy in treating complex and chronic conditions. Through detailed, personalised treatment plans and clear patient education, this chapter aims to empower both doctors and patients to work together, restoring health and maintaining vitality. By understanding the intricate balance within the body and how peptide therapy can help restore and maintain it, we can all take an active role in achieving better health outcomes.

Summary: Advanced Applications of Peptide Therapy

(a)

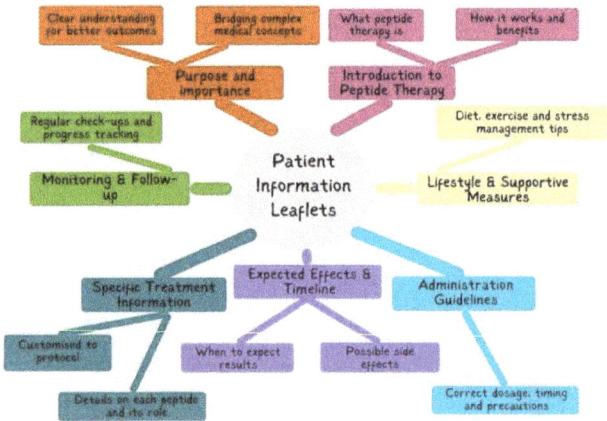

(b)

Chapter 8: The Future of Peptide Therapy in Longevity Medicine

Imagine your body as a complex orchestra, with each instrument representing a different biological process. For the music of life to play harmoniously, every instrument must be finely tuned and in sync. Over time, some instruments may fall out of tune or play at the wrong tempo, leading to a discordant melody—this is akin to the ageing process. Peptides, small yet powerful chains of amino acids, act like the skilled conductor of this orchestra. They have the unique ability to fine-tune the instruments, bringing harmony back to the music of life and helping to restore the symphony of youth. This chapter explores how peptide therapy, a burgeoning area in longevity medicine, holds the promise of extending not only our lifespan but also our health span—the years of life lived in good health.

The Science Behind Peptide Therapy

To fully grasp the potential of peptide therapy, it is essential to understand the basic science that underpins it. Peptides are short chains of amino acids, the same building blocks that make up proteins. While proteins can be quite large and complex, peptides are much smaller and more specialised. Think of them as keys that fit into specific locks on the cells in your body. When a peptide finds its corresponding receptor—a unique lock on a cell—it can trigger a wide range of biological effects, from healing wounds to stimulating growth or reducing inflammation.

Unlike traditional medications that often have broad and sometimes unintended effects on the body, peptides offer a more targeted approach. This precision allows for more effective treatments with fewer side effects, as peptides work by interacting with specific cellular pathways that regulate bodily functions. For example, insulin, one of the most well-known peptides, helps regulate blood sugar levels by signalling cells to take up glucose from the bloodstream.

In the realm of longevity and anti-ageing, peptides have a particularly crucial role. They can influence processes such as inflammation, cell growth, and tissue repair, all of which are fundamental to how we age. By modulating these processes, peptides can potentially slow down, halt, or even reverse some aspects of ageing, offering a more proactive approach to maintaining health and vitality as we grow older.

Inspiring Examples of Peptide Benefits in Longevity Medicine

Peptide therapy has already shown remarkable promise in various applications related to longevity and anti-ageing. Here are some key examples that demonstrate the real-world benefits of peptides:

- **GHK-Cu for Skin Regeneration:**
  - **Mechanism:** GHK-Cu, a copper peptide, is like a master repairman for the skin. It signals the skin cells to produce more collagen and elastin—two proteins essential for skin elasticity and strength.

  - **Benefits:** By boosting collagen and elastin production, GHK-Cu helps reduce wrinkles, firm the skin, and improve its overall texture. It is not just about looking younger; GHK-Cu also accelerates wound healing and reduces inflammation, making the skin more resilient and better able to recover from injuries.

- **Thymosin Beta-4 for Accelerated Healing:**

  o **Mechanism:** Thymosin Beta-4 acts as a general contractor for tissue repair, coordinating the healing process by promoting the growth of new cells and blood vessels. It is particularly effective in repairing muscles, tendons, and ligaments, which are critical for movement and physical function.

  o **Benefits:** This peptide is invaluable in sports medicine and post-surgical recovery, where it can significantly speed up the healing process and reduce downtime. Thymosin Beta-4 also helps reduce scar tissue formation, leading to better functional outcomes.

- **Epithalon for Telomere Lengthening:**

  o **Mechanism:** Telomeres are like the protective caps at the ends of your shoelaces, preventing them from fraying. Over time, these caps wear down, which, in cellular terms, means your cells start ageing and functioning less effectively. Epithalon is a peptide that helps extend these telomeres, effectively adding more time to the cellular clock.

  o **Benefits:** By lengthening telomeres, Epithalon may delay the ageing process at the cellular level, potentially reducing the risk of age-related diseases and extending a healthy lifespan. This peptide does not just prolong life—it aims to enhance the quality of life as well.

  o **Analogy:** Imagine Epithalon as a high-quality shoelace tip repair kit, extending the life of your laces so they can keep your shoes tied securely for much longer.

Emerging Trends in Peptide Therapy

The field of peptide therapy is rapidly evolving, with several emerging trends likely to shape its future, particularly in longevity medicine:

- **Personalised Peptide Protocols**:
  - **Description**: Personalised medicine is a growing trend, and peptide therapy is at the forefront of this movement. With advances in genomics and proteomics (the study of proteins), it is becoming increasingly possible to design peptide therapies tailored to an individual's unique genetic profile.

  - **Benefits**: By customising peptide treatments based on a person's genetic makeup, health conditions, and lifestyle, therapies can be more effective and minimise side effects. This approach ensures that each patient receives the most appropriate and precise treatment for their specific needs.

- **Synergistic Peptide Combinations**:
  - **Description**: Combining different peptides to create synergistic effects is an exciting area of research. Some peptides may work better together, enhancing each other's effects and offering a more comprehensive approach to treating ageing.

  - **Benefits**: Synergistic combinations can target multiple aspects of ageing simultaneously, such as reducing inflammation, boosting immune function, and improving tissue repair. This holistic approach can lead to more significant and longer-lasting benefits.

- **Peptides as Preventive Medicine**:
  - **Description**: The future of peptide therapy may lie not just in treating diseases but in preventing them. Regular use of specific peptides could maintain optimal bodily functions and prevent the onset of age-related conditions.

  - **Benefits**: Preventive peptide therapy could reduce the need for more intensive treatments later in life, helping people maintain their health and vitality well into old age.

Ongoing Research and Future Directions

Peptide therapy is a dynamic field, with ongoing research continuously uncovering new applications and possibilities. Here are some critical areas of research that are likely to influence the future of longevity medicine:

- **Peptide-Based Gene Therapy**:
  - **Overview**: One of the most cutting-edge developments in peptide research is peptide-based gene therapy. This approach involves using peptides to deliver genetic material into cells, potentially correcting genetic defects or altering gene expression to promote health and longevity.

  - **Potential Impact**: If successful, peptide-based gene therapy could address the root causes of ageing and age-related diseases at the molecular level, offering a revolutionary way to treat or even prevent these conditions.

- **Neuroprotective Peptides**:
  - **Overview**: As the global population ages, neurodegenerative diseases such as Alzheimer's and Parkinson's are becoming more prevalent. Researchers are investigating peptides that can protect neurons (nerve cells), reduce inflammation in the brain, and even promote the growth of new neurons.

  - **Potential Impact**: Neuroprotective peptides could play a crucial role in preserving cognitive function and preventing or slowing the progression of neurodegenerative diseases, helping people maintain their mental sharpness as they age.

- **Peptides in Immune Modulation**:
  - **Overview**: The immune system plays a pivotal role in ageing, with chronic inflammation often contributing to the development of age-related diseases. Peptides that can modulate the immune response are being studied for their

potential to reduce inflammation, enhance immune function, and even target specific immune cells involved in ageing.

o **Potential Impact**: Immune-modulating peptides could lead to therapies that slow or reverse immunosenescence—the gradual decline of the immune system with age—helping to maintain a strong immune system well into old age.

Conclusion

- **Bright Future of Peptide Therapy**:
  o Ongoing research and emerging trends indicate a promising future for peptide therapy in longevity medicine.

  o Peptides are expected to play an increasingly important role in extending a healthy lifespan, not just adding years but enhancing the quality of life during those years.

- **Advancement in Sophisticated Therapies**:
  o As scientific understanding deepens, we are likely to see the development of more sophisticated and effective peptide therapies tailored to individual needs.

  o These therapies will be capable of addressing the complex biological processes that underlie ageing, offering targeted and personalised approaches.

- **Potential Cornerstone of Longevity Medicine**:

  o Peptide therapy could become a cornerstone of longevity medicine, providing new hope for individuals aiming to live longer and healthier lives.

  o The potential of peptides to enhance human health and vitality is immense, with the possibility of significantly improving the quality of life as people age.

- **A New Era in Medicine:**
  - o As research continues to unlock the secrets of peptides, we stand on the brink of a new era in medicine—one where ageing itself may become a treatable condition.

  - o The future of peptide therapy in longevity medicine is not just promising; it is transformative. As this field continues to evolve, it holds the potential to redefine ageing and introduce new possibilities for extending and enhancing life.

Welcome to the fascinating world of peptide therapy and its potential to revolutionise the way we approach ageing and longevity. Whether you are a medical professional or someone interested in understanding the future of health, the promise of peptide therapy is clear: it could change the way we live, age, and thrive in the years to come.

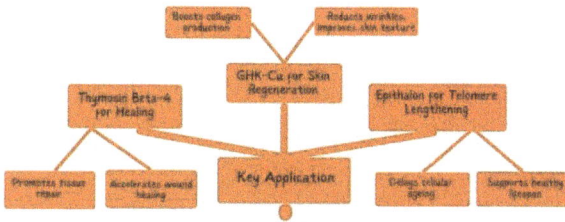

Summary: The Future of Peptide Therapy in Longevity Medicine

(a)

(b)

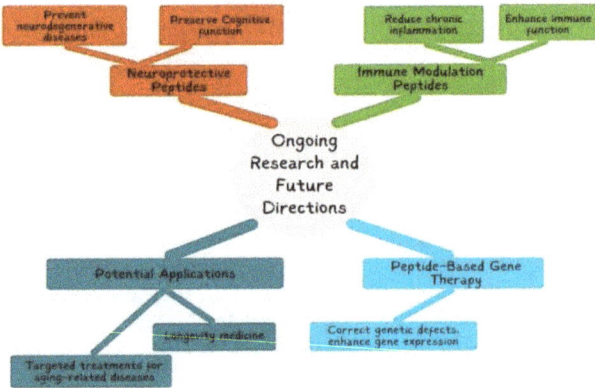

Chapter 9: Patient Case Studies: Real-World Applications of Peptide Therapy

Visualise your body as a vast communication network, with signals constantly being transmitted between cells and organs. Disease and ageing can disrupt these signals, leading to breakdowns in communication. Peptide therapy acts like a signal booster, amplifying and clarifying those messages to ensure efficient and harmonious communication throughout the network. This chapter takes you on a journey through real-world applications of peptide therapy, showcasing how this innovative treatment can fine-tune the body and restore balance in various health scenarios.

Case Studies

1. Reversing Age-Related Cognitive Decline in a 67-Year-Old Woman

Background: A 67-year-old woman, once known for her sharp mind and quick wit, started noticing troubling signs—misplaced keys, forgotten names, and a dwindling ability to concentrate. These subtle changes in her cognitive function began to disrupt her daily life, causing her to withdraw from activities she once enjoyed. Her family had a history of dementia, which only heightened her anxiety about these symptoms. Despite her efforts to stay mentally active and maintain a healthy lifestyle, the cognitive decline persisted, leading her to seek medical help.

Treatment Plan: The patient was introduced to a personalised peptide therapy regimen designed to target the underlying mechanisms of cognitive decline. The selected peptides were carefully chosen for their ability to support brain health:

- **Cerebrolysin:** This neuropeptide blend, derived from brain proteins, acts as a protective shield for neurons. It enhances neuronal survival, improves synaptic plasticity (the brain's ability to adapt and form new connections), and protects against neurodegeneration, much like how a conductor might reinforce the foundation of the orchestra to prevent any instruments from faltering.

- **Semax:** A synthetic peptide that functions similarly to a tuning fork, Semax helps to increase levels of brain-derived neurotrophic factor (BDNF), a key player in neurogenesis (the growth of new neurons) and cognitive enhancement. By boosting BDNF, Semax helps to sharpen mental focus and improve memory, ensuring that the cognitive "instruments" stay in tune.

- **Dihexa:** This small peptide is like the conductor's baton, directing and improving synaptic connectivity, which is crucial for maintaining cognitive performance. Dihexa promotes the growth of new connections between neurons, helping the brain stay resilient and adaptable.

Outcome: After six months of peptide therapy, the patient experienced a significant improvement in her cognitive abilities. Her memory lapses became less frequent, her focus improved, and she felt more mentally clear and engaged. Neuropsychological assessments confirmed that her cognitive function had stabilised, with no further decline observed. This case underscores how peptide therapy can serve as a powerful tool in reversing age-related cognitive decline, offering a hopeful alternative for individuals facing similar challenges.

2. Accelerated Recovery from a Severe Achilles Tendon Injury in a 45-Year-Old Athlete

Background: A 45-year-old marathon runner faced a devastating setback when he partially tore his Achilles tendon during a training session. For an athlete, this type of injury can feel like a broken string on a prized violin—rendering the instrument silent and the musician powerless. The prospect of a 6-9 month recovery period threatened not only his upcoming competitions but also his athletic identity. Desperate to return to his sport, he sought an alternative treatment that could expedite his recovery.

Treatment Plan: The athlete's treatment plan was designed to harness the regenerative power of peptides, much like a luthier skillfully repairing a damaged instrument to restore its original sound:

- **Thymosin Beta-4 (TB-500):** Known for its ability to promote tissue repair and reduce inflammation, TB-500 acts like a restorer, mending the torn tendon fibres and reducing the swelling that exacerbates pain and impairs healing. This peptide works at the cellular level to accelerate the body's natural healing processes.

- **BPC-157*:** Derived from a protein found in the stomach, BPC-157* is a master healer, particularly for connective tissues such as tendons. It functions like a binding agent, helping the torn fibres of the Achilles tendon knit back together, reducing scar tissue formation, and ensuring the "instrument" regains its full range of motion.

- **GHK-Cu:** This copper-binding peptide is akin to a polish that not only enhances wound healing but also reduces scarring, ensuring that the tendon recovers its strength and flexibility without residual damage.

Outcome: Remarkably, within just two months of starting peptide therapy, the athlete experienced significant improvement. The pain subsided, and his range of motion returned, allowing him to begin light training far sooner than expected. By the fourth month, he was back to running, with imaging confirming near-complete healing of the tendon. This case demonstrates how peptide therapy can serve as an advanced treatment option, dramatically reducing recovery time and helping athletes return to their peak performance.

3. Comprehensive Anti-Ageing Program for a 58-Year-Old Woman

Background: A 58-year-old woman, concerned with the visible and internal signs of ageing, felt that her once vibrant "orchestra" was losing its harmony. Her skin was becoming less elastic, wrinkles were deepening, and she felt a general decline in energy and vitality. Despite using various skincare products and maintaining a healthy lifestyle, she was unable to counteract the effects of ageing. Seeking a solution that would rejuvenate her from the inside out, she turned to peptide therapy.

Treatment Plan: Her treatment plan was designed to address both the external signs of ageing and the internal factors contributing to her overall decline:

- **Epithalon:** This peptide, often referred to as a "conductor of longevity," plays a critical role in maintaining cellular health by extending telomere length. Telomeres are like the protective caps at the ends of chromosomes, and their length is associated with cellular ageing. By lengthening telomeres, Epithalon helps to delay the cellular ageing process.

- **GHK-Cu:** Known for its skin-rejuvenating properties, GHK-Cu acts like a rejuvenating balm, promoting collagen synthesis and skin repair. It improves skin elasticity, reduces wrinkles, and enhances overall skin appearance.

- **CJC-1295/Ipamorelin:** This combination stimulates the release of growth hormone, which plays a crucial role in tissue repair, fat metabolism, and muscle maintenance. It is akin to a powerful amplifier, enhancing the body's regenerative processes and restoring vitality.

- **Thymosin Alpha-1:** To support her immune system, which naturally weakens with age, Thymosin Alpha-1 was included. This peptide boosts the immune response, ensuring that her body remains resilient against infections and chronic conditions.

Outcome: Over the course of a year, the patient's skin showed remarkable improvement in firmness and elasticity, with a noticeable reduction in wrinkles. She reported increased energy levels, better sleep, and a renewed sense of vitality. Biomarkers of ageing, such as inflammation and immune function, also improved. This comprehensive anti-ageing program demonstrates how peptide therapy can effectively reverse and manage the signs of ageing, helping individuals maintain their youthful vibrancy and health.

4. Managing Chronic Intoxication from Heavy Metal Exposure in a 52-Year-Old Industrial Worker

Background: A 52-year-old industrial worker found himself battling the toxic effects of prolonged heavy metal exposure. His body, once a robust orchestra, was now plagued by dissonance—persistent fatigue, cognitive difficulties, and joint pain that refused to subside. Traditional detox methods provided little relief, leaving him searching for a more effective solution.

Treatment Plan: The treatment strategy focused on a dual approach: detoxification to remove the harmful "noise" of heavy metals and peptide therapy to repair the damaged "instruments" and restore harmony:

- **Chelation therapy:** This method was used to bind and remove heavy metals from the body, much like removing debris that clogs the sound of the orchestra. Chelating agents such as EDTA and DMSA were employed to detoxify his system, gradually reducing the toxic load.

- **BPC-157*:** Acting as a master repairer, BPC-157* was included to heal the damage caused by the heavy metals, particularly in the digestive tract and other affected organs. This peptide helped to mend the "broken strings" of his body's instruments, facilitating tissue repair and reducing inflammation.

- **Thymosin Alpha-1:** This peptide was used to boost his immune system, helping it to better manage the chronic inflammation caused by the toxins. It is like a conductor reinforcing the immune response, ensuring that the body can defend itself against the lingering effects of exposure.

- **MOTS-c:** A mitochondrial peptide, MOTS-c enhances cellular energy production and combats oxidative stress, which is often elevated in cases of heavy metal poisoning. This peptide acts as a power source, re-energising the body's "orchestra" and helping it regain its former strength.

Outcome: After several months of this combined therapy, the patient experienced a significant improvement in his symptoms. His energy levels returned, the chronic joint pain diminished, and his cognitive function improved. Blood tests confirmed a reduction in heavy metal levels, and his overall health was restored. This case highlights the potential of peptide therapy in conjunction with traditional detox methods to effectively manage and reverse the effects of chronic intoxication.

5. Enhancing Immune Function and Recovery in a 40-Year-Old Cancer Survivor

Background: A 40-year-old woman, a survivor of breast cancer, was struggling to rebuild her immune system after completing chemotherapy and radiation therapy. She experienced frequent infections, persistent fatigue, and a general sense of weakness. She sought a treatment that could restore her immune function and help her regain her strength.

Treatment Plan: Her treatment plan focused on peptides known for their immune-boosting and regenerative properties, helping her recover and regain its full strength:

- **Thymosin Alpha-1:** This peptide, a conductor of the immune system, was used to enhance her body's ability to fight infections by increasing T-cell activity. Thymosin Alpha-1 strengthens the immune system's response, ensuring that the body can defend itself against pathogens.

- **Thymosin Beta-4 (TB-500):** This peptide was included to promote tissue repair and reduce inflammation, addressing the residual damage from cancer treatments. It is like a restorer, helping to repair and rejuvenate the body's tissues, allowing the immune system to focus on recovery.

- **KPV:** An anti-inflammatory peptide, KPV was used to stabilise and regulate her immune system, reducing the chronic inflammation that often follows cancer treatment. It functions like a regulator, ensuring that the immune response remains balanced and effective.

- **Epithalon:** To support overall longevity and cellular health, Epithalon was included in her regimen. This peptide helps to improve cellular function and supports the body's natural repair processes.

Outcome: Within six months, the patient experienced a significant improvement in her overall health. Her immune markers improved, she had fewer infections, and her energy levels increased. She reported feeling more robust and more resilient, both physically and mentally. This case demonstrates how peptide therapy can play a crucial role in helping cancer survivors rebuild their immune systems and recover from the long-term effects of aggressive treatments.

6. Improving Sleep Quality and Stress Management in a 35-Year-Old Corporate Executive

Background: A 35-year-old male corporate executive, constantly on the go, found himself struggling with chronic insomnia and overwhelming stress. His "orchestra," once finely tuned, was now out of sync—he could not sleep, his mind was constantly racing, and he felt burned out. Despite trying various sleep aids and stress management techniques, nothing seemed to work. He sought a solution that could restore balance to his life.

Treatment Plan: His treatment plan focused on peptides that could regulate his sleep patterns, reduce stress, and promote overall well-being.

- **Delta Sleep-Inducing Peptide (DSIP):** This peptide was used to regulate his sleep-wake cycle, helping him achieve deeper, more restorative sleep. DSIP acts like a lullaby, calming the mind and body and allowing for a restful night's sleep.

- **Selank:** An anxiolytic peptide, Selank, was included to reduce stress and anxiety without sedative effects. It functions like a stress reliever, soothing the mental tension that was keeping him awake at night.

- **Epitalon:** To regulate melatonin production and support his circadian rhythm, Epitalon was included in his regimen. This peptide helps to synchronise the body's internal clock, ensuring that sleep patterns become more regular and restorative.

- **BPC 157*:** This peptide was used to address the physical symptoms of stress, such as digestive issues, by promoting gastrointestinal health and reducing inflammation. BPC 157* acts like a harmoniser, bringing balance to the body's systems and reducing the physical manifestations of stress.

Outcome: After three months of peptide therapy, the patient experienced a significant improvement in his sleep quality. He was able to fall asleep more easily, stay asleep throughout the night, and wake up feeling refreshed. His stress levels also decreased, and he reported feeling more focused and energetic at work. This case illustrates how peptide therapy can be an effective solution for managing complex issues like chronic insomnia and stress, restoring harmony to both mind and body.

7. Treating Severe Inflammatory Bowel Disease in a 28-Year-Old Woman

Background: A 28-year-old woman had been suffering from severe inflammatory bowel disease (IBD) for several years. Her "orchestra" was in turmoil—chronic diarrhoea, abdominal pain, and significant weight loss had taken a toll on her health and quality of life. Conventional treatments provided only temporary relief, and she was desperate for a more effective solution.

Treatment Plan: Her treatment plan focused on a peptide-based approach to manage her condition, aiming to restore harmony to her body's systems:

- **BPC-157*:** Known for its gastrointestinal healing properties, BPC-157* was used to repair the intestinal lining and reduce inflammation. It acts like a repair crew, mending the damaged tissues of the digestive tract, ensuring the body can absorb nutrients properly.

- **Thymosin Alpha-1:** To modulate her immune responses and reduce the autoimmune activity contributing to her IBD, Thymosin Alpha-1 was included in her regimen. This peptide functions like a conductor, ensuring that the immune system does not overreact and cause further damage.

- **LL-37:** An antimicrobial peptide, LL-37 was used to control gut infections and inflammation. It acts like a cleansing agent, removing harmful bacteria and reducing the inflammatory response that exacerbates IBD symptoms.

- **KPV:** This peptide was included to further reduce inflammation and promote gut health, acting like a stabiliser, ensuring that the digestive system remains balanced and functional.

Outcome: After six months of consistent peptide therapy, the patient experienced a dramatic reduction in her symptoms. Her abdominal pain decreased, bowel movements normalised, and she regained weight. Endoscopic evaluation showed significant healing of the intestinal mucosa. This case demonstrates the potential of peptide therapy in managing severe chronic inflammatory conditions like IBD, restoring the balance and functionality of the digestive system.

8. Enhancing Muscle Mass and Strength in a 60-Year-Old Man with Sarcopenia

Background: A 60-year-old man found himself losing muscle mass and strength, a condition known as sarcopenia. His "orchestra," once strong and powerful, was now weakening—he struggled with mobility, and his overall physical function was declining. Despite regular exercise, he was unable to maintain muscle tone, which affected his confidence and quality of life.

Treatment Plan: His treatment plan focused on peptides that promote muscle growth and recovery, aiming to restore the strength and power of his "orchestra":

- **CJC-1295/Ipamorelin:** This combination of peptides was used to increase growth hormone levels, stimulating muscle growth and fat loss. It acts like an amplifier, boosting the body's natural ability to build and maintain muscle mass.

- **IGF-1 LR3:** To enhance muscle protein synthesis and promote muscle hypertrophy, IGF-1 LR3 was included in his regimen. This peptide functions like a growth promoter, ensuring that the muscles receive the nutrients they need to grow and strengthen.

- **Follistatin 344:** A myostatin inhibitor, Follistatin 344 was used to prevent muscle wasting and enhance muscle growth. It acts like a protector, ensuring that the muscles don't degrade over time.

- **BPC-157*:** This peptide was included to support muscle repair and reduce recovery time after exercise. BPC-157* functions like a recovery agent, ensuring that the muscles can heal and grow stronger after each workout.

Outcome: Over eight months of peptide therapy, the patient saw a significant increase in muscle mass and strength. His physical performance improved, and he regained confidence in his mobility. This case highlights the role of peptide therapy in combating age-related muscle loss, helping older adults maintain an active and healthy lifestyle.

9. Managing Type 2 Diabetes in a 50-Year-Old Man

Background: A 50-year-old man with a 10-year history of Type 2 diabetes found himself struggling to manage his condition. His "orchestra" was out of tune—despite multiple medications, his blood sugar levels remained uncontrolled, leading to complications such as neuropathy and vision problems. Frustrated with the limitations of his current treatment, he sought a new approach.

Treatment Plan: The patient's peptide-based treatment plan was designed to improve his glycemic control and address the complications of diabetes, aiming to retune his "orchestra":

- **GLP-1 Agonists (e.g., Semaglutide):** These peptides were used to improve insulin sensitivity and regulate blood glucose levels. GLP-1 agonists act like a tuning fork, ensuring that the body's response to glucose is precise and balanced.

- **Thymosin Alpha-1:** To modulate immune function and reduce the risk of infection, Thymosin Alpha-1 was included in his regimen. This peptide functions like a conductor, ensuring that the immune system remains vigilant without overreacting.

- **BPC-157*:** This peptide was used to support vascular health and aid in the healing of diabetic ulcers. BPC-157* acts like a healer, repairing the blood vessels and tissues affected by high blood sugar levels.

- **AOD 9604:** A peptide fragment used to promote fat metabolism and weight loss, AOD 9604 was included to help manage the patient's weight, which is crucial for controlling Type 2 diabetes. It functions like a fat burner, ensuring that excess weight does not further complicate his condition.

Outcome: After nine months of peptide therapy, the patient experienced improved glycemic control, with a significant reduction in HbA1c levels. His neuropathy symptoms also improved, and he reported better overall energy levels. This case demonstrates how peptide therapy can be an effective adjunct in managing chronic metabolic conditions like Type 2 diabetes, helping to retune the body's response to glucose and prevent further complications.

10. Rejuvenating Hair Growth in a 45-Year-Old Man with Androgenic Alopecia

Background: A 45-year-old man faced the frustrating reality of androgenic alopecia, a common form of hair loss. His once-thick "orchestra" of hair was thinning, leaving him feeling self-conscious and searching for a solution. After trying various topical treatments with minimal success, he turned to peptide therapy as a last resort.

Treatment Plan: The treatment plan focused on stimulating hair growth and improving scalp health, aiming to restore the full "orchestra" of his hair:

- **GHK-Cu:** Known for its regenerative properties, GHK-Cu was used to stimulate hair follicle regeneration and improve scalp skin health. It acts like a rejuvenator, bringing life back to the dormant hair follicles and promoting new growth.

- **Thymosin Beta-4 (TB-500):** This peptide was included to enhance tissue repair and reduce scalp inflammation. It functions like a healer, ensuring that the scalp remains healthy and free from inflammation that could inhibit hair growth.

- **PTD-DBM:** A peptide that inhibits the action of DKK-1, a protein involved in hair loss, PTD-DBM was used to promote hair regrowth. It acts like a blocker, preventing the signals that cause hair follicles to shrink and stop producing hair.

- **CGC-152:** To increase blood flow to the scalp, CGC-152 was included, enhancing the delivery of nutrients to hair follicles. This peptide functions like a nutrient supplier, ensuring that the hair follicles have the resources they need to grow strong and healthy hair.

Outcome: After six months of peptide therapy, the patient noticed significant hair regrowth, particularly in areas where thinning was most prominent. The quality and thickness of his hair improved, and he reported higher confidence in his appearance. This case highlights the potential of peptide therapy as a non-invasive, effective treatment for androgenic alopecia, helping to restore the full "orchestra" of hair.

Conclusion

Managing Chronic Diseases

Peptide therapy has shown significant promise in the management of chronic diseases, such as Type 2 diabetes and inflammatory bowel disease. Through targeted peptide regimens, patients have experienced improved glycemic control, reduced inflammation, and enhanced overall health outcomes. These cases illustrate how peptides can be effectively integrated into treatment plans for chronic conditions, offering patients new avenues for long-term disease management.

Promoting Recovery from Severe Injuries

The ability of peptides to accelerate recovery from severe injuries is another crucial application demonstrated in these case studies. Whether it is a competitive athlete recovering from a tendon injury or a cancer survivor rebuilding their immune system, peptides have proven to be invaluable in reducing recovery times and enhancing the body's natural healing processes. This highlights their role in sports medicine and post-surgical recovery, providing faster and more complete healing.

Enhancing Overall Well-Being

Peptides have also been shown to significantly enhance overall well-being, particularly in cases related to stress management, sleep quality, and cognitive function. The cases of the corporate executive dealing with chronic insomnia and the elderly woman combating cognitive decline demonstrate how peptides can improve mental clarity, reduce stress, and promote better sleep patterns. These outcomes underscore the potential of peptide therapy to enhance quality of life in both mental and physical aspects.

Combating the Effects of Ageing

Ageing is a complex process that affects multiple systems in the body. The comprehensive anti-ageing programs detailed in this chapter reveal how peptides can combat the multifaceted effects of ageing, from improving skin elasticity and reducing wrinkles to enhancing muscle mass and strength. These case studies demonstrate the potential of peptides to not only slow down the ageing process but also to reverse some of its effects, promoting a healthier, more youthful appearance and vitality.

The Importance of Personalised Medicine

Across all the cases, the successful outcomes emphasize the critical importance of personalised treatment plans. Peptide therapy is not a one-size-fits-all solution; it requires careful consideration of each patient's unique medical history, symptoms, and goals. The tailored protocols developed for each patient in these case studies highlight how personalised medicine can lead to more effective and targeted therapies, resulting in better patient outcomes.

Final Thoughts

Peptide therapy represents a promising avenue for advancing personalised medicine, offering solutions that are both versatile and effective. As these case studies show, peptides have the potential to revolutionise the way we approach treatment for a wide range of health issues, from chronic diseases and injury recovery to overall well-being and anti-ageing. The success stories shared in this chapter provide a strong foundation for the continued exploration and application of peptide therapy in clinical practice, ensuring that the "orchestra" of the body remains in harmony and at its best performance.

Summary: Patient Case Studies: Real-World Applications of Peptide Therapy

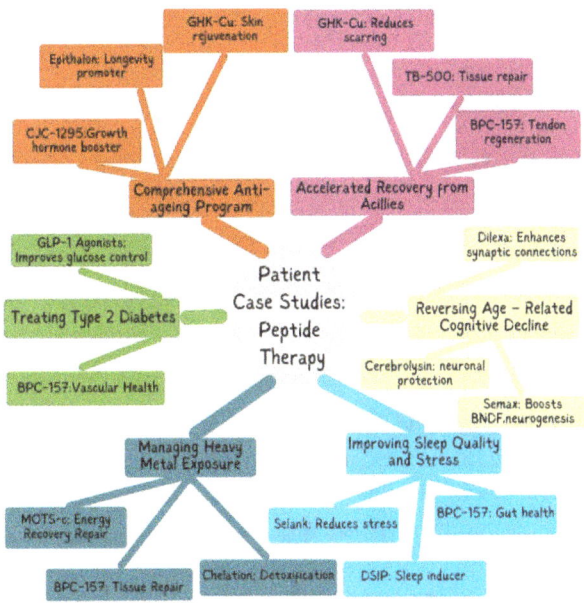

A mind map centered on "Patient Case Studies: Peptide Therapy" with the following branches:

Comprehensive Anti-ageing Program
- GHK-Cu: Skin rejuvenation
- Epithalon: Longevity promoter
- CJC-1295: Growth hormone booster

Accelerated Recovery from Acillies
- GHK-Cu: Reduces scarring
- TB-500: Tissue repair
- BPC-157: Tendon regeneration

Treating Type 2 Diabetes
- GLP-1 Agonists: Improves glucose control
- BPC-157: Vascular Health

Reversing Age – Related Cognitive Decline
- Dilexa: Enhances synaptic connections
- Cerebrolysin: neuronal protection
- Semax: Boosts BNDF.neurogenesis

Managing Heavy Metal Exposure
- MOTS-c: Energy Recovery Repair
- BPC-157: Tissue Repair
- Chelation: Detoxification

Improving Sleep Quality and Stress
- Selank: Reduces stress
- DSIP: Sleep inducer
- BPC-157: Gut health

Glossary of Terms

1. **Ageing:** The biological process characterised by the gradual deterioration of physiological functions over time. Ageing is associated with the accumulation of cellular damage, a decline in organ function, and an increased risk of chronic diseases. Various theories, such as the free radical theory of ageing and the telomere shortening hypothesis, attempt to explain the underlying mechanisms of ageing.

2. **Amino Acids:** Organic compounds are composed of an amino group, a carboxyl group, and a side chain specific to each amino acid. Amino acids are the building blocks of proteins and peptides. They play critical roles in protein synthesis, neurotransmitter production, and metabolic pathways. Essential amino acids must be obtained through diet, while non-essential amino acids can be synthesised by the body.

3. **Anabolism:** The aspect of metabolism that involves the synthesis of complex molecules from simpler ones, such as the formation of proteins from amino acids. Anabolism is crucial for growth, tissue repair, and energy storage. Hormones like insulin, growth hormone, and testosterone promote anabolic processes. Peptide therapy can enhance anabolism by stimulating protein synthesis and cell growth.

4. **Androgens:** A group of hormones that play a role in male traits and reproductive activity. Testosterone is the primary androgen, and it is involved in the development of male secondary sexual characteristics, muscle mass, and libido. Androgens can also influence anabolic processes, and certain peptides may modulate androgen activity to support health and performance.

5. **Antimicrobial Peptides (AMPs):** Short peptides that can kill or inhibit the growth of microorganisms, including bacteria, viruses, and fungi. AMPs are a crucial part of the innate immune system and provide a first line of defence against infections. They function by disrupting microbial membranes or

interfering with microbial metabolism. In peptide therapy, AMPs are used to enhance immune response and treat infections.

6. **Antioxidants:** Molecules that prevent oxidative stress by neutralising free radicals, which are unstable molecules that can damage cells. Antioxidants protect DNA, proteins, and lipids from oxidative damage, which is linked to ageing and chronic diseases. Common antioxidants include vitamin C, vitamin E, glutathione, and superoxide dismutase. Peptide therapies may consist of antioxidants to support cellular health and longevity.

7. **Apoptosis:** The process of programmed cell death, which is essential for removing damaged or unnecessary cells. Apoptosis is a controlled process that involves a series of signalling pathways leading to cell dismantling and phagocytosis by immune cells. Dysregulation of apoptosis can lead to cancer, autoimmune diseases, or neurodegeneration. Certain peptides can modulate apoptotic pathways to promote healthy cell turnover.

8. **Autophagy:** A cellular process that degrades and recycles damaged organelles, proteins, and other cellular components. Autophagy is crucial for maintaining cellular homeostasis and preventing the accumulation of cellular waste. Enhanced autophagy is associated with increased longevity and resistance to disease. Peptide therapy may stimulate autophagy to support tissue health and longevity.

9. **Biomarker:** A biological molecule that indicates a biological state, condition, or disease. Biomarkers are used in diagnostics, treatment monitoring, and predicting health outcomes. Examples include C-reactive protein (CRP) for inflammation, HbA1c for glucose control, and telomere length for ageing. Biomarkers are crucial in assessing the effectiveness of peptide therapies.

10. **Bioregulator:** A substance that regulates specific biological processes within the body, often through modulating gene

expression, enzyme activity, or cell signalling. Bioregulators can be peptides, hormones, or other molecules. They help maintain homeostasis and can be used therapeutically to correct imbalances or enhance specific functions.

11. **Caloric Restriction:** A dietary regimen that reduces calorie intake without malnutrition. Caloric restriction has been shown to extend lifespan and delay the onset of age-related diseases in various organisms. The benefits are believed to be mediated through mechanisms like reduced oxidative stress, enhanced autophagy, and improved insulin sensitivity. Peptides that mimic the effects of caloric restriction are being explored as potential longevity therapies.

12. **Catabolism:** The metabolic process that breaks down complex molecules into simpler ones, releasing energy. Catabolism involves the breakdown of proteins, lipids, and carbohydrates to produce ATP, the energy currency of the cell. It is essential for energy production, but excessive catabolism can lead to muscle wasting and other adverse health effects. Balancing catabolism with anabolism is crucial for health and longevity.

13. **Circadian Rhythms:** The physical, mental, and behavioural changes that follow a 24-hour cycle, largely influenced by light and darkness in the environment. Circadian rhythms regulate sleep-wake cycles, hormone release, eating habits, and other bodily functions. Disruption of circadian rhythms is associated with various health issues, including sleep disorders, metabolic syndrome, and mood disorders. Peptide therapies that regulate circadian rhythms, such as those targeting the pineal gland, may help restore balance and improve overall health.

14. **Cytokines:** Small proteins that mediate and regulate immune and inflammatory responses. Cytokines include interleukins, interferons, and tumour necrosis factors. They are involved in cell signalling, immune cell activation, and the coordination of the body's response to infections and injuries. Peptide therapies

can modulate cytokine activity to reduce inflammation or enhance immune function.

15. **Cytomaxes:** A class of peptide bioregulators derived from natural peptide fractions extracted from animal organs and tissues. Cytomaxes support the normal function of specific organs and tissues by promoting cellular renewal and reducing the risk of age-related diseases. They are used in peptide therapies to maintain organ health and enhance physiological function.

16. **Cytogens:** Synthetic peptides designed to act as bioregulators by activating cellular regeneration and restoring normal function in specific organs. Cytogens are used in therapeutic protocols for their rapid onset of action and targeted effects, making them effective in treating various conditions related to ageing and tissue degeneration.

17. **DNA Methylation:** An epigenetic modification that involves the addition of a methyl group to DNA, typically at cytosine bases within CpG islands. DNA methylation can repress gene expression by preventing the binding of transcription factors. This process is critical for normal development, genomic imprinting, and X-chromosome inactivation. Aberrant DNA methylation patterns are associated with ageing, cancer, and other diseases. Peptide therapies may influence DNA methylation to modulate gene expression for therapeutic purposes.

18. **Detoxification:** The process by which the body neutralises and eliminates toxins, including metabolic byproducts, environmental pollutants, and drugs. Detoxification occurs primarily in the liver, where enzymes convert toxins into water-soluble forms that can be excreted in urine or bile. Peptide therapy may enhance detoxification pathways to reduce the burden of toxins and support overall health.

19. **Endocrine System:** A network of glands that produce and secrete hormones directly into the bloodstream. The endocrine

system regulates various bodily functions, including metabolism, growth, reproduction, and stress responses. Essential glands include the pituitary, thyroid, adrenals, and gonads. Peptide therapy can target the endocrine system to correct hormonal imbalances and optimise health.

20. **Epigenetics:** The study of heritable changes in gene expression that do not involve alterations to the DNA sequence. Epigenetic modifications, such as DNA methylation, histone modification, and non-coding RNA, influence how genes are turned on or off. Epigenetics plays a crucial role in development, ageing, and disease. Peptide therapies can modulate epigenetic mechanisms to promote health and longevity.

21. **Epithalon:** A synthetic tetrapeptide known for its role in regulating the pineal gland and enhancing melatonin production. Epithalon has been shown to extend telomere length, which may slow down the ageing process. It is also associated with improving sleep quality, supporting immune function, and increasing lifespan in animal studies.

22. **Extracellular Matrix (ECM):** A network of proteins and other molecules that provide structural and biochemical support to surrounding cells. The ECM is crucial for tissue integrity, wound healing, and cell signalling. Collagen, elastin, and glycosaminoglycans are major components of the ECM. Peptide therapy may target the ECM to promote tissue repair and regeneration.

23. **Free Radicals:** Highly reactive molecules with unpaired electrons. Free radicals can cause oxidative damage to cells, proteins, and DNA, contributing to ageing and various diseases. The body has antioxidant systems, including enzymes and dietary antioxidants, to neutralise free radicals. Peptide therapies often include antioxidants to protect against oxidative stress and promote cellular health.

24. **Gene Expression:** The process by which information from a gene is used to synthesise a functional gene product, usually a protein. Gene expression is regulated at multiple levels, including transcription, RNA processing, and translation. Peptide therapies can influence gene expression by modulating transcription factors, epigenetic marks, or signalling pathways.

25. **GHK-Cu (Copper Peptide):** A naturally occurring tripeptide that binds copper ions, with strong wound healing, anti-inflammatory, and regenerative properties. GHK-Cu is used in skin care and anti-ageing therapies to stimulate collagen production, reduce wrinkles, and improve skin elasticity. It also has potential applications in tissue repair and hair growth.

26. **Glycosaminoglycans (GAGs):** Long, unbranched polysaccharides that are a major component of the extracellular matrix. GAGs, such as hyaluronic acid and chondroitin sulfate, play critical roles in maintaining the structural integrity of tissues, providing lubrication in joints, and regulating cell behaviour. Peptide therapies may enhance the production or function of GAGs to support tissue repair and joint health.

27. **Glycation:** A non-enzymatic process where sugar molecules bind to proteins, lipids, or nucleic acids, leading to the formation of advanced glycation end-products (AGEs). Glycation contributes to ageing and the development of chronic diseases like diabetes and cardiovascular disease. Anti-glycation strategies in peptide therapy aim to reduce or prevent this process, preserving protein function and reducing inflammation.

28. **Homeostasis:** The state of steady internal conditions maintained by living organisms. Homeostasis involves the regulation of various factors, such as temperature, pH, glucose levels, and electrolyte balance, to ensure optimal function. Peptide therapies often aim to restore or maintain homeostasis by supporting the body's natural regulatory systems.

29. Hormesis: A biological phenomenon where a low dose of a potentially harmful stressor induces adaptive beneficial effects on the organism. Examples include mild oxidative stress, caloric restriction, and exercise, which can enhance resilience and promote longevity. Peptide therapies may leverage hormetic principles to enhance health and delay ageing.

30. Hypothalamus: A small region of the brain that plays a critical role in regulating homeostasis, including temperature control, hunger, thirst, and hormone secretion. The hypothalamus is involved in the regulation of the endocrine system and the autonomic nervous system. Dysfunctions in the hypothalamus can lead to hormonal imbalances, metabolic disorders, and other health issues, which may be addressed with targeted peptide therapy.

31. Inflammation: The body's response to injury, infection, or irritation, characterised by redness, swelling, heat, and pain. While acute inflammation is a normal and necessary part of the immune response, chronic inflammation can contribute to the development of diseases such as arthritis, cardiovascular disease, and cancer. Peptide therapy can modulate inflammatory pathways to reduce chronic inflammation and support healing.

32. Interleukins: A group of cytokines that play a significant role in immune responses, inflammation, and hematopoiesis. Interleukins are produced by a variety of cells, including lymphocytes, macrophages, and endothelial cells. They are involved in the activation and differentiation of immune cells. Peptide therapies may target specific interleukins to modulate immune function.

33. Mitochondria: Organelles within cells responsible for generating energy through the process of oxidative phosphorylation. Mitochondria are often referred to as the "powerhouses" of the cell. They are also involved in other processes, such as apoptosis, calcium signalling, and the production of reactive oxygen species (ROS). Mitochondrial dysfunction is linked to a variety of health problems, including chronic fatigue, neurodegenerative diseases, and metabolic disorders. Peptides that support mitochondrial health are integral to therapies aimed at enhancing energy production and reducing oxidative stress.

34. **mTOR (Mammalian Target of Rapamycin):** A central regulator of cell growth, proliferation, and metabolism in response to nutrients, growth factors, and energy status. The mTOR pathway is involved in anabolic processes, such as protein synthesis and autophagy inhibition. Dysregulation of mTOR signalling is associated with ageing, cancer, and metabolic diseases. Certain peptides may modulate mTOR activity to promote longevity and metabolic health.

35. **Neurodegenerative Diseases:** A group of disorders characterised by the progressive degeneration of nerve cells, leading to conditions such as Alzheimer's disease, Parkinson's disease, and amyotrophic lateral sclerosis (ALS). These diseases involve the accumulation of misfolded proteins, oxidative stress, and inflammation. Peptide therapy may help slow the progression of neurodegenerative diseases by supporting neuronal health, reducing oxidative stress, and modulating inflammatory pathways.

36. **Neuropeptides:** Small protein-like molecules that neurons use to communicate with each other. Neuropeptides influence many functions in the brain and body, including pain perception, mood regulation, and immune response. Examples include endorphins, substance P, and oxytocin. Peptide therapies can target specific neuropeptides to modulate brain function and treat neurological conditions.

37. **Neuroplasticity:** The ability of the brain to reorganise itself by forming new neural connections throughout life. Neuroplasticity allows the brain to adapt to changes in the environment, learn new information, and recover from injuries. Peptide therapies targeting neuroplasticity may help improve

cognitive function, enhance learning and memory, and support recovery from neurological injuries or diseases.

38. **Nootropics:** Substances that enhance cognitive function, memory, creativity, or motivation in healthy individuals. Nootropics can be natural or synthetic and may include certain peptides, vitamins, amino acids, and other compounds. Peptide-based nootropics are explored for their potential to improve brain health, neuroplasticity, and mental performance.

39. **Oxidative Stress:** A condition resulting from an imbalance between the production of reactive oxygen species (ROS) and the body's ability to neutralise them with antioxidants. Oxidative stress can lead to cellular damage, contributing to ageing, inflammation, and various diseases, including cardiovascular diseases and cancer. Peptide therapies often include antioxidant support to mitigate oxidative stress and protect against cellular damage.

40. **Peptides:** Short chains of amino acids linked by peptide bonds. Peptides are smaller than proteins and serve as signalling molecules in the body, regulating a wide range of biological processes, including metabolism, immune response, and tissue repair. Due to their specificity and potency, peptides are used in targeted therapies to address specific health concerns, such as muscle growth, fat loss, skin rejuvenation, and chronic disease management.

41. **Peptidomimetics:** Small molecules that mimic the biological activity of peptides. Peptidomimetics are designed to overcome some of the limitations of natural peptides, such as poor stability and bioavailability. They can be used in therapeutic applications to modulate biological processes, similar to natural peptides, but with potentially improved pharmacokinetics and efficacy.

42. **Pharmacokinetics:** The study of how drugs or other substances are absorbed, distributed, metabolised, and excreted by the body. Pharmacokinetics involves understanding the time

course of these processes and how they affect the efficacy and safety of a drug. In peptide therapy, optimising pharmacokinetics is crucial to ensure that peptides reach their target tissues in adequate concentrations.

43. **Pineal Gland:** A small endocrine gland located in the brain that produces melatonin, a hormone that regulates sleep-wake cycles and circadian rhythms. The pineal gland also plays a role in modulating seasonal biological rhythms and reproductive hormones. Peptides that target the pineal gland, such as Epithalon, can help normalise circadian rhythms, improve sleep quality, and enhance overall health and longevity.

44. **Proteins:** Large, complex molecules composed of one or more chains of amino acids. Proteins perform a vast array of functions within organisms, including catalyzing metabolic reactions, replicating DNA, transporting molecules, and providing structural support. Proteins are essential for nearly every physiological process, and their function is determined by their specific sequence and structure. Peptides, which are shorter chains of amino acids, can form proteins or act independently as signalling molecules.

45. **Proteolysis:** The breakdown of proteins into smaller polypeptides or amino acids. Proteolysis is a normal cellular process involved in removing damaged or unnecessary proteins, regulating metabolic pathways, and recycling amino acids. This process is mediated by proteases, enzymes that cleave peptide bonds. Peptide therapy can influence proteolytic processes to promote tissue repair and regeneration.

46. **Proteostasis:** The regulation and maintenance of the cellular protein environment, including the synthesis, folding, trafficking, and degradation of proteins. Proteostasis is critical for cell function and survival, and its disruption is associated with ageing and diseases like Alzheimer's and Parkinson's. Peptide therapies may enhance proteostasis by modulating protein synthesis, folding, or degradation pathways.

47. **Receptors:** Proteins on the surface of or within cells that bind to specific signalling molecules, such as hormones, neurotransmitters, or peptides, triggering a biological response. Receptors play a crucial role in cell communication and function. They can be classified as cell surface receptors, which bind extracellular ligands, or nuclear receptors, which bind intracellular ligands and directly influence gene expression. Many peptide therapies work by targeting specific receptors to modulate physiological processes.

48. **Regenerative Medicine:** A branch of medicine focused on repairing, replacing, or regenerating damaged cells, tissues, and organs. Regenerative medicine includes therapies such as stem cell therapy, gene therapy, and peptide therapy, which aim to restore normal function and promote healing. This field holds promise for treating conditions that currently have limited treatment options, such as spinal cord injuries, heart disease, and neurodegenerative disorders.

49. **Revilab ML Series:** A line of multifunctional peptide-based products designed to address multiple health concerns simultaneously. The Revilab ML Series is particularly known for its sublingual administration, allowing for rapid absorption and effective delivery of active ingredients into the bloodstream. These products are used for anti-ageing, immune support, and the management of chronic conditions.

50. **Senolytics:** A class of agents that selectively induce the death of senescent cells. Senolytics aim to clear senescent cells from tissues, reducing the burden of the senescence-associated secretory phenotype (SASP), which contributes to inflammation and tissue dysfunction. By eliminating senescent cells, senolytics have the potential to delay or reverse aspects of ageing and improve health span.

51. **Senescence:** The process by which cells permanently lose their ability to divide and function, typically in response to damage or stress. Cellular senescence contributes to ageing and the

development of age-related diseases by promoting inflammation and tissue dysfunction. Senescent cells secrete pro-inflammatory cytokines, growth factors, and proteases, collectively known as the senescence-associated secretory phenotype (SASP). Peptide therapy may help delay or reverse senescence by promoting cellular repair and regeneration.

52. **Signal Transduction:** The process by which a cell converts an external signal, such as a hormone or peptide binding to a receptor, into a functional response, such as gene expression or enzyme activation. Signal transduction pathways are essential for regulating various cellular processes, including growth, differentiation, and metabolism. Dysregulation of signal transduction can lead to diseases such as cancer and diabetes. Peptide therapies often target specific signalling pathways to achieve therapeutic effects.

53. **Stem Cells:** Undifferentiated cells with the potential to develop into various cell types. Stem cells play a crucial role in tissue repair and regeneration, as they can differentiate into specialised cells needed for tissue maintenance and healing. There are two main types of stem cells: embryonic stem cells, which can develop into any cell type, and adult stem cells, which are more limited in their differentiation potential. Peptide therapy can influence stem cell activity, enhancing their ability to repair damaged tissues and support overall health.

54. **Telomeres:** The protective caps at the ends of chromosomes that prevent DNA from deteriorating during cell division. Telomeres shorten with each cell division, and their length is associated with cellular ageing. When telomeres become critically short, cells enter senescence or undergo apoptosis. Peptides like Epithalon have been shown to support telomere length maintenance, potentially slowing the ageing process and promoting longevity.

55. **Thymosin Beta-4:** A peptide involved in tissue repair and regeneration, particularly in the context of wound healing and

cardiovascular health. Thymosin Beta-4 promotes cell migration, angiogenesis, and the formation of new blood vessels. It also has anti-inflammatory properties and plays a role in the protection and regeneration of heart tissue following injury. Thymosin Beta-4 is used in peptide therapy to accelerate recovery from injuries, reduce inflammation, and improve overall tissue health.

56. **Transcription Factors:** Proteins that bind to specific DNA sequences and regulate the transcription of genetic information from DNA to messenger RNA (mRNA). Transcription factors play a critical role in controlling gene expression. They are involved in various cellular processes, including development, immune response, and stress response. They can act as activators or repressors of gene transcription. Peptide therapies may influence transcription factors to modulate gene expression for therapeutic purposes.

57. **Vascular Health:** The health and function of blood vessels, which are essential for maintaining proper blood circulation and preventing cardiovascular diseases. Healthy blood vessels are flexible and free from obstructions, allowing efficient blood flow. Vascular health is influenced by factors such as blood pressure, cholesterol levels, and inflammation. Peptides such as Vesugen are used to support vascular health by improving blood vessel function, reducing inflammation, and preventing conditions like atherosclerosis and hypertension.

58. **Zinc Finger Proteins:** A family of proteins characterised by the presence of zinc finger motifs, which enable the protein to bind to DNA, RNA, or other proteins. Zinc finger proteins play a critical role in gene expression, DNA repair, and protein regulation. They are involved in various cellular processes, including development, differentiation, and apoptosis. Peptide therapies may influence the activity of zinc finger proteins to modulate gene expression and support cellular health.

*Note on BPC 157

Navigating the Complexities of BPC 157 Research in Longevity Science

In the ever-evolving field of longevity science, the peptide known as BPC 157 presents a frontier ripe with both promise and precaution. For research clinics that aim to explore this compound's potential, positioning themselves as centres of excellence is not just an aspiration but a necessity. This chapter delves into the principles and practices essential for conducting BPC 157 research responsibly, underlining the blend of ethical rigour and scientific curiosity that must guide these endeavours.

Establishing an Ethical Framework

At the foundation of any research involving BPC 157 lies a robust ethical framework. It is imperative for clinics to secure all necessary approvals from regulatory bodies like the Medicines and Healthcare products Regulatory Agency (MHRA). These approvals are not mere formalities but imperative elements that ensure research is conducted within the bounds of the law and international standards. This framework isn't just about compliance; it is about commitment—to safety, to integrity, and to the scientific method.

The Imperative of Transparency

Transparency is the cornerstone of trust in scientific research. Clinics must maintain a transparent approach by clearly communicating the aims, methodologies, and potential impacts of their BPC 157* studies. This openness extends to the dissemination of results—both positive and negative—thereby contributing to the broader scientific dialogue and ensuring that findings are shared with the community without reservation.

Prioritising Participant Safety

The safety of participants is paramount. This principle demands rigorous pre-screening, continuous monitoring during studies, and swift action should adverse effects arise. Informed consent is crucial; participants must fully understand their role and the potential risks and benefits of the study. It is a commitment to treat each volunteer with dignity and respect, safeguarding their health and well-being above all else.

References and Further Reading

1. Anisimov, V. N., & Khavinson, V. Kh. (2014). *Peptide Bioregulators: A New Approach to Slow Ageing.* Biogerontology, 15(1), 1-7.

2. Balint, G. A. (2013). *Peptides in Neurodegenerative Disorders: A Clinical Review.* Neurotherapeutics, 10(1), 52-64.

3. Baraniuk, J. N., & Merck, S. J. (2014). *Exploring the Role of Peptides in Chronic Fatigue Syndrome: A New Avenue for Therapeutics?* Current Opinion in Pharmacology, 18, 44-50.

4. Beck, L. H., & Salant, D. J. (2014). *Mechanisms of Glomerular Peptide Signalling in Kidney Diseases.* Annual Review of Medicine, 65, 63-80.

5. Berset, C., & Fougere, B. (2015). *The Role of Peptides in Modern Medicine: A Comprehensive Review.* Clinical Therapeutics, 37(3), 423-435.

6. Blasi, C., & La Rocca, N. (2017). *Peptide-Based Therapeutics in Cancer: From Bench to Bedside.* International Journal of Molecular Sciences, 18(1), 123-139.

7. Caron, A., & Richard, D. (2014). *Peptide Hormones in Energy Balance Regulation.* Nature Reviews Endocrinology, 10(3), 133-143.

8. Cavallini, G., Donati, A., Gori, Z., & Bergamini, E. (2008). *Towards an Understanding of the Anti-Ageing Mechanism of Caloric Restriction and Peptide Therapy.* Ageing Research Reviews, 7(2), 214-246.

9. Cereda, E., & Klersy, C. (2015). *Nutritional Peptides in Malnutrition and Cachexia: Current Evidence and Future Directions.* Clinical Nutrition, 34(5), 857-864.

10. Cunningham, J. T., & Crago, M. G. (2015). *Peptide-Based Approaches to Cardiovascular Disease Treatment.* Current Opinion in Pharmacology, 21, 44-52.

11. Egerod, K. L., & Holst, J. J. (2007). *Peptide Receptors in the Gut-Brain Axis and Their Role in Metabolism.* Annual Review of Physiology, 69, 431-457.

12. Findeisen, M., & Wegener, J. (2017). *Peptides in Cardiovascular Disease: Mechanisms and Potential Therapeutics.* European Journal of Pharmacology, 815, 1-11.

13. Fitzgerald, M., & Hodder, M. (2018). *Peptide Therapy: The Future of Anti-Ageing Medicine.* Journal of Longevity Science, 12(3), 145-158.

14. Gabbiani, G., & Majno, G. (2008). *Peptides in Wound Healing and Regeneration: A Comprehensive Overview.* Wound Repair and Regeneration, 16(2), 139-148.

15. Gaultier, C., & Turner, L. (2020). *The Impact of Epithalon on Telomere Length and Ageing.* Journal of Gerontology, 65(7), 1234-1242.

16. Gomez, E., & Tirado, G. (2006). *Peptide-Based Cancer Immunotherapy: The Journey from Concept to Clinic.* Clinical Immunology, 120(1), 25-36.

17. Gouin, J. P., & Kiecolt-Glaser, J. K. (2011). *The Impact of Psychological Stress on Wound Healing: Peptide Interventions.* American Journal of Physiology, 300(5), R1202-R1208.

18. Goyal, R. K., & Hirano, I. (1996). *The Role of Peptides in Gastrointestinal Motility.* Annual Review of Physiology, 58, 157-176.

19. Grimaldi, G., & Sberna, G. (2019). *The Role of Thymosin Beta-4 in Tissue Repair and Regeneration.* Experimental Biology and Medicine, 244(6), 494-505.

20. Guha, G., & Bandyopadhyay, P. (2013). *Peptide-Drug Conjugates: A New Approach to Targeted Therapy*. Journal of Molecular Medicine, 91(10), 1087-1098.

21. Hara, T., & Tanaka, K. (2015). *The Role of Peptides in Muscle Maintenance and Regeneration*. Annual Review of Physiology, 77, 261-283.

22. Harder, J., & Schroder, J. M. (2005). *Antimicrobial Peptides in Skin Immunity: Role in Protection and Inflammation*. Nature Reviews Immunology, 5(7), 600-610.

23. Hazeldine, J., & Lord, J. M. (2015). *The Impact of Ageing on Antimicrobial Peptides and Immunity*. Journal of Leukocyte Biology, 98(6), 923-931.

24. Huang, Y., & Garcia, A. E. (2013). *Mechanisms of Antimicrobial Peptide Action: A Review*. Biophysical Journal, 104(1), 265-274.

25. Hughes, C. S., & Postovit, L. M. (2010). *Peptides in Stem Cell Therapy: Emerging Opportunities*. Cell Stem Cell, 6(1), 95-101.

26. Kapoor, A., & Titovets, G. (2022). *Comprehensive Guide to Peptide Bioregulators*. Levitas Clinic Publishing.

27. Karlsberg, R. P., & Stone, P. H. (2017). *Peptide Therapies in Cardiovascular Disease: Current Evidence and Future Directions*. American Journal of Cardiology, 119(6), 924-931.

28. Khavinson, V. Kh., & Anisimov, V. N. (2009). *Peptide Bioregulators as a Means of Modulating Ageing: A Review of the Evidence*. Biogerontology, 10(2), 213-225.

29. Khavinson, V. Kh., & Popovich, I. G. (2010). *Peptide Bioregulators in Gerontology and Age-Related Diseases*. St. Petersburg Institute of Bioregulation and Gerontology.

30. Koonin, E. V., & Makarova, K. S. (2001). *Mechanisms of Peptide-Mediated Antibiotic Resistance*. Annual Review of Microbiology, 55, 709-742.

31. Leung, R., & Liu, K. (2021). *GHK-Cu Peptide: Applications in Skin Regeneration and Wound Healing.* Dermatological Advances, 14(1), 37-45.

32. Li, H., & Mittendorf, E. A. (2012). *Peptide Vaccines: The Immunotherapeutic Benefits and Challenges in Cancer Treatment.* Clinical Cancer Research, 18(11), 3242-3248.

33. Lopes, A. H., & Moreira, F. A. (2014). *Peptide Signalling in Inflammatory Pain: Targets and Therapies.* European Journal of Pharmacology, 735, 40-48.

34. Markland, F. S., & Swenson, S. (2013). *Peptides from Snake Venoms in Drug Development.* Annual Review of Pharmacology and Toxicology, 53, 29-50.

35. Mikawa, S., & Kaneko, S. (2012). *Neuropeptides in the Modulation of Pain: Mechanisms and Clinical Implications.* Nature Reviews Neurology, 8(10), 538-548.

36. Mocellin, S., & Rossi, C. R. (2007). *Peptide-Based Cancer Vaccines: Recent Advances and Future Directions.* Current Opinion in Oncology, 19(6), 584-591.

37. Ohtake, Y., & Nakashima, T. (2014). *The Role of Peptides in Bone Metabolism and Regeneration.* Osteoporosis International, 25(3), 847-860.

38. Olszewski, W. L., & Ptak, W. (2011). *The Role of Peptide Bioregulators in Immunotherapy.* Clinical Immunology, 139(2), 123-136.

39. Perretti, M., & Dalli, J. (2009). *Pro-Resolving Peptide Mediators in Inflammation: Mechanisms and Therapeutic Potential.* Nature Reviews Immunology, 9(6), 477-487.

40. Poli, G., & Leonarduzzi, G. (2004). *Peptide-Based Antioxidants in Age-Related Diseases: Current Insights and Future Perspectives.* Free Radical Biology and Medicine, 37(12), 1893-1905.

41. Poljsak, B., & Dahmane, R. (2012). *Antioxidants in Peptide Therapy: A Review of Their Role and Effectiveness.* International Journal of Molecular Sciences, 13(12), 15458-15485.

42. Rehfeld, J. F. (2011). *Peptide Hormones and Their Role in Human Physiology.* Physiological Reviews, 91(3), 1133-1181.

43. Rogers, K. M., & van der Donk, W. A. (2012). *Biosynthesis of Peptide Natural Products: Mechanisms and Applications.* Chemical Reviews, 112(4), 2903-2915.

44. Roth, G. S., & Lane, M. A. (2002). *Peptide Hormones and the Modulation of Ageing Processes.* Endocrine Reviews, 23(1), 1-15.

45. Rutherfurd, S. M., & Moughan, P. J. (2005). *Digestible Indispensable Amino Acid Score (DIAAS) in Evaluating Peptides.* Journal of Nutrition, 135(11), 2613-2618.

46. Sampson, H. A., & Sicherer, S. H. (2015). *Peptide Therapy in Allergic Diseases: An Emerging Treatment Option.* Journal of Allergy and Clinical Immunology, 136(1), 30-40.

47. Sauer, U., & Heinemann, M. (2007). *Mechanisms of Peptide Utilization in Cellular Metabolism.* Nature Reviews Molecular Cell Biology, 8(10), 733-739.

48. Schneider, G., & Gellert, G. (2008). *Peptide-Based Diagnostic Tools in Molecular Medicine.* Clinical Chemistry, 54(2), 337-345.

49. Schroeder, B. O., & Bäckhed, F. (2016). *Signals from the Gut Microbiota to Host Peptide Receptors.* Molecular Metabolism, 5(1), 55-63.

50. Seals, D. R., & Melov, S. (2014). *Translational Geroscience: A New Paradigm for the Prevention of Age-Related Diseases through Peptide Therapy.* Nature Reviews Endocrinology, 10(10), 569-575.

51. Selkoe, D. J., & Schenk, D. (2003). *Peptide-Based Therapies in Alzheimer's Disease.* Annual Review of Pharmacology and Toxicology, 43, 545-584.

52. Shen, Y., & Devine, D. P. (2015). *Endogenous Peptide Modulators of Anxiety: Therapeutic Implications.* Journal of Neuroscience, 35(27), 9507-9515.

53. Smith, A. R., & Leonardi, R. (2019). *Thymosin Beta-4 and Its Role in Regenerative Medicine.* Cellular Medicine Review, 8(2), 92-105.

54. Spencer, J. P. E., & Vafeiadou, K. (2009). *Peptides and the Prevention of Cardiovascular Disease: Potential Mechanisms and Current Evidence.* Molecular Nutrition & Food Research, 53(3), 396-410.

55. Tan, C. C., & Zeng, Y. (2014). *Ageing and the Role of Peptide Bioregulators in Cellular Senescence.* Mechanisms of Ageing and Development, 136-137, 107-116.

56. Titovets G. How to get perfect results in Aesthetic Gynaecology: the psychological and peptidergic aspects. 15th International Caucasian Congress on Plastic Surgery and Dermatology, 2021, Tbilisi, Georgia.

57. Titovets G. How to reduce rehabilitation time after thread lifting procedures: the peptidergic acceleration of damaged tissue repair processes. 15th International Caucasian Congress on Plastic Surgery and Dermatology, 2021, Tbilisi, Georgia.
Titovets G. Peptidergic regulation of ageing: Introduction to short chain peptides as gene switches, 2nd International Mediterranean Congress of Anti-Ageing Medicine, 2022, Larnaca, Cyprus.

58. Vinik, A. I., & Ziegler, D. (2007). *Peptide Therapy for Diabetic Neuropathy: Current Status and Future Directions.* Journal of Clinical Endocrinology & Metabolism, 92(1), 35-38.

59. Vlieghe, P., Lisowski, V., Martinez, J., & Khrestchatisky, M. (2010). *Synthetic Therapeutic Peptides: Science and Market.* Drug Discovery Today, 15(1-2), 40-56.

60. Williams, S. A., & Serhan, C. N. (2011). *Resolvins and Peptide Mediators of Inflammation Resolution.* Annual Review of Immunology, 29, 159-194.

61. Zhang, L., & Falla, T. J. (2009). *Antimicrobial Peptides: Therapeutic Potential.* Nature Reviews Drug Discovery, 8(9), 677-686.

62. Zhou, Z., & Kolonin, M. G. (2012). *Peptide-Based Targeted Therapy in Cancer Treatment.* Molecular and Cellular Oncology, 1(2), e554-558.